A Journey into Open Science and Research Transparency in Psychology

A Journey into Open Science and Research Transparency in Psychology introduces the open science movement from psychology through a narrative that integrates song lyrics, national parks, and concerns about diversity, social justice, and sustainability. Along the way, readers receive practical guidance on how to plan and share their research, matching the ideals of scientific transparency.

This book considers all the fundamental topics related to the open science movement, including: (a) causes of and responses to the Replication Crisis, (b) crowdsourcing and meta-science research, (c) preregistration, (d) statistical approaches, (e) questionable research practices, (f) research and publication ethics, (g) connections to career topics, (h) finding open science resources, (i) how open science initiatives promote diverse, just, and sustainable outcomes, and (j) the path moving forward. Each topic is introduced using terminology and language aimed at intermediate-level college students who have completed research methods courses. But the book invites all readers to reconsider their research approach and join the Scientific Revolution 2.0. Each chapter describes the associated content and includes exercises intended to help readers plan, conduct, and share their research.

This short book is intended as a supplemental text for research methods courses or just a fun and informative exploration of the fundamental topics associated with the Replication Crisis in psychology and the resulting movement to increase scientific transparency in methods.

Jon Grahe is Professor of psychology and department chair at Pacific Lutheran University, USA. Other roles include managing executive editor of *The Journal of Social Psychology*, president of the Western Psychological Association, and former president of Psi Chi, the International Honor Society. He also led the design and administration of the Collaborative Replications and Education Project (CREP) and the Emerging Adulthood Measured at Multiple Institutions (EAMMi2) project, among other undergraduate crowd projects.

"In 2013, Jon Grahe convinced Mark Brandt and I of his dream to come along on his journey to teach replication projects across universities, which now has become widely known as the Collaborative Replication and Education Project. Jon's book provides an excellent introduction to the principles that convinced Mark and I to come along on his journey: to make the world a better place via high-quality research that does justice to the human condition. Jon provides an in-depth discussion that is partly historical, partly forward-looking in his characteristically story-telling way, drawing from his own journey covering research practices, statistics, ethics, writing, diversity, and even career advice. Read Jon's book to understand why this will become the go-to introduction to open science."

– **Hans Rocha IJzerman** *is an Associate Professor at Université Grenoble Alpes, France, and author of* Heartwarming: How Our Inner Thermostat Made Us Human

"Who knew that rock music, national parks, and replication could be woven together into an accessible narrative that introduces the reader to open science principles and practices? Grahe provides an effective introduction to research rigor and transparency with a perfect blend of conceptual instruction, concrete examples, and learn-by-doing. After completing A Journey into Open Science and Research Transparency in Psychology readers won't just know about open science, they'll be doing it themselves!"

– **Brian Nosek** *is co-founder and executive director of the Center for Open Science. He is also a Professor in the department of psychology at the University of Virginia, USA*

"Grahe's visionary textbook leverages the potential for psychology undergraduates not only to learn about research methods but also to do valuable projects themselves. Armed with cutting-edge tools for open, transparent, and reproducible research, a history of the recent upheavals in science, and an understanding of the relation between scientific and societal values, students will be prepared for the conceptual and technical scientific challenges of the future."

– **Barbara Spellman,** *Professor of psychology, University of Virginia, USA*

A Journey into Open Science and Research Transparency in Psychology

Jon Grahe

NEW YORK AND LONDON

First published 2022
by Routledge
605 Third Avenue, New York, NY 10158

and by Routledge
2 Park Square, Milton Park, Abingdon, Oxon, OX14 4RN

Routledge is an imprint of the Taylor & Francis Group, an informa business

Library of Congress Cataloging-in-Publication Data
A catalog record for this book has been requested

ISBN: 978-0-367-47159-0 (hbk)
ISBN: 978-0-367-46459-2 (pbk)
ISBN: 978-1-003-03385-1 (ebk)

Typeset in Palatino
by Apex CoVantage, LLC

Access the Support Material: www.routledge.com/9780367464592

CONTENTS

Preface *vi*

Chapter 1 **A Replication Crisis: Responses Benefit Personal Workflow 1**

Chapter 2 **Go Forth and Replicate: Making Methods for Others 14**

Chapter 3 **Preregistered: Determining Answers to Decisions Before They Happen 32**

Chapter 4 **Decision Heavyweights: Drawing Inference With Confidence 47**

Chapter 5 **Ode to p-Hacking: Making Decisions Before They Happen 65**

Chapter 6 **You Can't Plagiarize Yourself: Avoiding Errors With Ethical Writing 81**

Chapter 7 **Becoming a Second Stringer: Why Good People Do Replication Science 96**

Chapter 8 **Open Science Alphabet: Learning to Read 109**

Chapter 9 **Progress: Open Science Promotes Diverse, Just, and Sustainable Outcomes 129**

Chapter 10 **Scientific Transparency: A Theme for a Movement 148**

Index *166*

PREFACE

This book invites readers to complete a journey through open science. Within the pages, I describe the open science movement and associated initiatives and outcomes at an introductory level aimed at supplementing student learning as they complete research methods and capstone courses. In fact, the book itself is a bit of a personal capstone project for me. Beginning in 2010, I was trying to convince my peers that we could do better research by pooling our resources and conducting research collectively. My efforts yielded little success until the Replication Crisis in 2011 jolted the field into making changes. At this point, the larger community's passion for replication coincided with my desire to create more and better research opportunities for undergraduates.

As I watched these events unfold, I quickly adopted the open science movement's goals as my own, though my own passion had been more about building a better science by becoming more efficient. I subsequently began championing undergraduate research opportunities and open science principles in every role and professional venue I could access. As I entered my next sabbatical, I wanted to share my experience and knowledge before transitioning my scholarship focus to something different. What came from that desire to share was the "Crisis Schmeisis Lecture or Music Tour," and this book transforms that experience into a narrative to introduce open science in a fun and informative way. The goal is not to replace the classroom text, which introduces basic methods, but rather to augment that material with the tools needed to present those methods transparently.

In this way, it is a capstone project for me. For students, capstone projects (also called senior research) represent the culminating educational experience that synthesizes (or attempts to) all the knowledge and learning to that point. An undergraduate capstone project ideally represents the summation and product of the student's experience. Though not every course or construct is included in the expansive project, the project comes from the collective experiences that make up that education.

In the same way, this book brings together multiple aspects of my personal and professional growth. The book includes four themes that propel the narrative forward. In the center are the open science movement and the goals of scientific transparency. However, my understanding of this content and my manifestation of its principles are influenced by the rest of me. As readers will soon recognize, the rest of me includes music, travel, and personal reflection. These extracurricular components guide me and keep me moving forward. By the time my sabbatical arrived, music, travel, and personal reflection were top priorities along with my need to produce scholarship from my work in the open science movement. And so I merged them. This integration of open science knowledge, music development, need for travel, and conviction to effortfully self-reflect represents a capstone to me. Here is a more expansive explanation for how and why these merged at this moment for me. Afterward, I will more specifically point to how they are reflected in the text.

To start with, this is the only psychology textbook, supplemental or otherwise, that I know of that is framed around the lyrics of a concept album. You can watch and/or listen to the songs on the Purrfect Second Stringers YouTube channel. I encourage readers to learn the songs and play along. However, the book does not require musical interest. The lyrics might be read as poetry or even just an outline of a talk. The lyrics are not presented within the text themselves, but the reader can access them online as part of the book's appendix.

However, the music was critical for me. My passion for making crowd projects of undergraduates made it impossible for me to continue playing music for fun. By the time of my sabbatical, I could barely play songs I knew well. Sabbatical is a time in which there is a bit more time, and so I wanted to re-engage with music. After the "Replication Crisis" song came in a moment of inspiration and developed quickly, I decided to use my love of open science and my vocation as a professor to help me relearn music. By committing to writing a concept album of songs about open science that could be used in a classroom, I was committing to playing music again. To force myself to learn more in the process, I wrote songs with increasing complexity or difficulty so that I had known challenges. In the end, the 10 songs challenge the guitar player through a range of keys and styles. Though my limitations are revealed in the live performances, the practice value of the 52 minutes of music was certainly evident to me.

An interest in travel does not make me unique. However, it influences this book, because while I was writing that concept album of "Songs to Inspire Scientific Transparency," I was also committed to visiting as many US national parks as possible. Because we live within a couple hours of two national parks (Olympic National Park and Mt. Rainier National Park) and because hiking and camping are among our most favorite pastimes, my wife and I buy the yearly pass that allows access to all national parks. It is a matter of simple logic for us. The cost of the pass is equal to visiting three different parks. If we already know we will visit two as the matter of a normal year, it is highly likely that we will have an opportunity to visit a third sometime during the year. It turns out to be a remarkably economical way to recreate if you don't have to pay for lodging.

Rather than trying to make the reader jealous, I would encourage anyone with the same privileged status to take advantage of these wild spaces for their own well-being. In any case, the only limits to travel normally are cost, work, and family commitments. When on sabbatical, work commitments are no longer an excuse, and a loving and supportive wife who shares an empty nest removes the family-commitment limits. Because cost is still a limit for me, I took advantage of work-related travel to allow me to also enjoy personal time.

For the Crisis Schmeisis Lecture and Music Tour, I gave talks and workshops wherever I could get in. It started when a school hired me to do a capstone workshop for them; I drove instead of riding in a plane and turned a three-day professional trip into a three-week professional road trip. Instead of just sharing open science with that one institution, I visited seven. Along the way, I drove through or visited seven national parks.

Over the course of the year, I traveled many times to professional conferences or meetings and, in each case, extended it to advance the tour. I traveled frugally, sleeping on couches or in the back of my van along the way. But I always sought another chance to bring the open science message forth. For example, when Psi Chi sent me to a conference in Texas, I rented a car and drove to Arkansas, turning a weekend trip into two weeks and visiting five institutions along the way. There were only two national parks on that trip, but they each made a lasting impact on me because they amplified my clarity regarding the third component of my personal development: personal reflection regarding diversity, social justice, and sustainability.

Across my adult life, I have tried to become a better person. What better means isn't always clear, but while serving on a committee in 2012, I found direction when my institution incorporated a commitment to advancing diversity, social justice, and sustainability in its long-range plan, called PLU2020. My direction came from the conversations that we had about them. The challenge was that there were proponents who argued the singular importance of each one compared to the other. For instance, they might argue that unless someone values diversity, there is no point to social justice or sustainability. Alternatively, without a livable planet, there is no possibility of social justice or diversity. The challenge was that all are equally important, and clarity came when all three were brought into a single focus. Rather than singularly considering diversity or social justice or sustainability, the institution would strive to acknowledge the trilogy as guideposts.

For me, this conflict and compromise represented the challenge and ideal in the larger social world. Certainly, the world includes a diverse array of values, including some that are quite contradictory to these, such as desires for autocracy, power, or greed. But for individuals driven toward a social good, these are values that are generally shared; "all people should be valued, all people should be treated equally, and decisions should be made that manage resources for the future." And yet there is inherent conflict between these value statements when enacted.

Following the challenge inherent in the institution's long-term plan, I concentrated on my own lens across these values. As I entered sabbatical, I still struggled with that clarity. As the fourth prong of my sabbatical plan, I decided to engage in deep personal reflection on the topics of diversity, social justice, and sustainability. Across the year, I accomplished this by reading books that were not part of my professional scope.

In keeping with the theme of the Crisis Schmeisis tour, I sought books within the context of national parks. At every national park visitor center bookstore, I looked for books that would help me expand my understanding of diversity, social justice, or sustainability. Often, I sought stories about or from individuals with minoritized backgrounds. Ideally, that book would put the person in an environmental conflict as well, such as *The Story of Luna* or *African American Women in the Old West*. In some instances, the book came from another event in my life, such as *What Does It Mean to Be White?* which was part of a reading group at my home institution and seemed applicable to my personal needs. In one case, the book (*Proud Shoes*) was chosen as a substitute because I could find no book store in any of the visitor center areas at Hot Springs National Park. When seeking something about the area, I stumbled across the book and found the story compelling. In each instance, I read the book and tried to place myself in the context of the story. As I drove through these beautiful spaces, I also recognized pain and hardship that I had not seen before. There were certainly stories of triumph, but there were more stories of pain, broken promises, and even murder in the name of civilization and manifest destiny.

Increasingly, I could see connections between the messages of promise from open science and the calls for a better society from the trifocal lens of diversity, social justice, and sustainability. This book represents my closing one chapter and moving to the next. As I hoped, the sabbatical plan brought forth clarity for the next phase of my professional career with a nice byproduct of reminding my fingers how to move on the neck of a guitar. It also allowed me to crystalize my message to researchers unfamiliar with open science principles. And so I share this book in an effort to expand the reach of this message. I hope that the reader finds the journey both pleasurable and engaging in addition to being informative. To achieve this, I incorporate all the components of my experience into each chapter.

As I mentioned earlier, each song was written following an intention that they would be successive and somewhat cumulative. The book is framed around these 10 songs. The title of each chapter matches the song title, and an "about the song" section follows the chapter abstract and objectives. These about-the-song sections explain the creative connection between the song and the chapter. There is also at least one "Crisis Schmeisis Book Review" in each chapter. These books were not read in any particular order, and they are not all directly or clearly tied to the content of each chapter. Instead, I chose books in each chapter that were the best fit after the book was completed. Throughout the book, the national parks are connected by the hypothetical Book Research Example about how many miles people hike at national parks. Additionally, the concepts of open science are often introduced through metaphors considering one or more national parks. Though this book is intended primarily for psychology or social science audiences, the national park context helps connect the topics without the need for any disciplinary expertise. At the end of each section are a few chapter exercises intended to help the reader further advance their own ongoing projects.

Across the book, readers will learn about the (Chapter 1) causes, consequences, and some responses to "The Replication Crisis." In Chapter 2, "Go Forth and Replicate" helps put the idea of replication into the context of science and suggests some ideas about doing it on a big scale. Chapter 3 explains all the issues that should be considered when preregistering research with a title intended to bring smiles; "Preregistered." Chapter 4 explains the competing approaches to statistical decisions with the metaphorically titled "Decision Heavyweights." In Chapter 5, "An Ode to p-Hacking" reflects on the many decisions researchers make and where they are represented in a manuscript with a title ironically lamenting questionable research practices. The title of Chapter 6, "You Can't Plagiarize Yourself," speaks to one aspect of research in ethics while the chapter explores ethics across the research process. Chapter 7 considers many aspects about careers generally while specifically explaining why people support open science initiatives, and in response to one critic's derisive label of replication scientists, it is titled "Becoming a Second Stringer." Chapter 8, "Open Science Alphabet," describes different examples of open science initiatives and suggests methods to keep up to date in the ever-changing landscape of trying to keep science transparent. Chapter 9 proclaims "Progress: Open Science Promotes Diverse, Just, and Sustainable Outcomes" after challenging the reader to engage in deep reflection on the topic. Chapter 10 envisions a future full of "Scientific Transparency" while similarly presenting realistic criticisms of the movement. My earnest hope is that the readers will follow this path with their own research and go forth to conduct transparent science.

ACKNOWLEDGMENTS

There are so many people who I should thank personally for helping me complete my own journey as I became a better scientist, reengaged music as a hobby, crystalized my worldview, and completed this book. There are really too many for me to list here, and any list will assuredly include some unintended omissions. The work of open science occurs in a very large community, with many people working on one project. And so I will keep my list very brief and limit it to people that helped me explicitly with the task of finishing this book and the Crisis Schmeisis album.

In terms of the music, I want to thank the members of Band of Waxx, Frank Murphy and Jeff Cason, who learned the songs for the debut of most of the songs. Along with Jeff, Andrew Franks and Amber Matteson joined me for the live debut of all the songs with The Purrfect Second Stringers in December 2019. That show was a highlight of this process because it was a test of the entertainment value of the music itself, and it was a complete delight. Amber also played the songs with me as my teaching apprentice for my statistics/methods students and I want to thank her for pressing me to teach her the songs, and subsequently some music. I am sure these songs are better because I knew that I had someone waiting for me to finish them.

When I sit down and think about the people who helped with the book, Amber is at the top of that list. As one of the few people who knew the songs as well as I did and who recently graduated with her BS in psychology after being my research assistant and teaching apprentice for methods courses, she offered a very good perspective as a reviewer. She also cowrote the "About the Song" boxes, since they were presented at WPA 2020. Finally, she wrote the R code that provided the images for Figure 4.2. In sum, her interest in the songs and open science were very helpful.

Another very helpful reviewer of the book was Leslie Cramblet Alvarez, whom I worked with on many open science initiatives and the Psi Chi board of directors. Because she was a coauthor on numerous prior projects and early adopter of undergraduate crowd projects, I could trust her to provide feedback that would advance my vision.

Another person who assisted with this project, Tiffany Williams, is a former student who earned an MA in library science and served as a personal research and editing assistant while I wrote the book. Having graduated in 2004, she was actually the student who asked the question that prompted me to write "You Can't Plagiarize Yourself," something I had forgotten until she texted me while reviewing the paper about how funny it was. I am very grateful to her for all the feedback and minor editing she offered during the process.

Though not directly related to writing the book, there are people who, for one reason or another were influential in helping me complete the book: Ronald Riggio, who said, "Wow! That's a great idea. You could do that for your career" in response to my ideas on how to fix psychology way back in 2009. Michelle Ceynar, whose friendship is eclipsed only by her commitment to her students and colleagues. Bobbie Spellman, for "un-rejecting" my commentary introducing the idea of crowdsourcing student projects. Jeffrey Spies for building the Open Science Framework as part of his dissertation and then co-founding the Center for Open Science with Brian Nosek, who has done so much himself worth acknowledging. Most notably, I am ever grateful for his recognition that I should work with Mark Brandt and Hans IJzerman to build the CREP.

And then for Martha Zlokovich, the staff at the Psi Chi Central Office, and those who served on the board who were willing to try out open science with me in many ways. Those early CREPers who helped when there wasn't enough help: Nicole Legate, Brady Wiggins, Lily Lazarevic, Cristina Baciu, and especially Jordan Wagge, who gave their time and energy to build an international project with zero operating budget. The people on the EAMMi2 planning committee who made that project work: Holly Chalk, Caitlyn Faas, and Joseph McFall. For both those projects, the dozens of faculty and hundreds of students who tried a new way to learn psychological methods. John Edlund, as Psi Chi research director, and Kelly Cuccolo, as first Network for International Collaborative Exchange director, who made that project work as envisioned by the Psi Chi board of directors.

There are too many PLU students who participated in one or more of these projects to list them all. But some students made extra efforts by working on projects after the term ended until they were complete or by serving as assistants to the larger projects. Nicole Bennet, Devin Bland, Katie Coddington, DeVere Dudley, Emily Fryberger, Halé Gervais, Katye Griswold, Samantha Henderson, Kaitlin Johnson, Hannah Juzeler, Andrew Nelson, Hailey Sandin, Meghan Schultz, Kelsey Serier, and Tiana Wamba all have my extra gratitude.

Last and maybe most importantly, I need to thank my wife, whose patience and understanding allowed me to keep my focus on these projects for many more hours than we expected. She allowed me the extra time I needed, and the emotional security, to engage in these projects and then to write this book. She was there supporting me whenever I needed her throughout this process and before. Now, she asked me not to put her in the acknowledgments, so I must ask the reader to proceed directly to reading the book and promise not to tell her that I thanked her.

CRISIS SCHMEISIS: SONGS TO INSPIRE SCIENTIFIC TRANSPARENCY

Background and Explanation: After recording two self-produced albums with My Name Aint Skip in 2010, the band split, and I stopped playing music except with my children. By 2016, they had found their own interests, and I had stopped actively engaging in music. Multiple attempts to write music or restart playing ended with failure. No doubt this was in part due to my immersion in open science activities. However, in September 2016, "The Replication Crisis" wrote itself as I walked my dog. Over the course of a week or so, all the verses were identified in my head, and I started thinking about picking up my guitar and figuring out a melody.

As anyone who has ignored their musical instrument for five years can tell you, even though I wanted to play chords, my fingers and hands were not following my brain's instructions. I ended up writing a much simpler song than I imagined just so I could play it. The situation remained unchanged with my single song and no plan until I found myself in a conversation with a group of open science enthusiasts who found my song idea compelling.

In response to some good-natured ribbing about making science into music, I decided to write an entire album about the movement. Following my approach in writing an earlier concept album, "MMiX: the Year" with My Name Aint Skip, I followed a set of guidelines to add structure to the process. The guidelines were as follows: (a) write the songs to follow an order which could accompany a methods course, (b) add something to each song that challenged me to be a better musician, and (c) write songs that would be interesting to people who were not involved in the movement.

The plan to start writing the album coincided with my sabbatical, so I incorporated the project into my plans to create better teaching material for open science. Finally, I had begun a journey to more deeply consider the intersections of diversity, social justice, and sustainability (DJS) four years earlier and devoted a portion of my time to reading and reflection. When a friend invited me to give a talk 1,500 miles away, I decided to make a road trip instead of fly and offer free open science talks or workshops to anyone interested along the way. This began the Crisis Schmeisis Open Science Musical or Talking Tour. It became important to then integrate travel into the other components of my sabbatical: music, open science, and DJS. Over the next 12 months, I represented open science almost 50 different times in 21 states in talks, meetings, workshops, and a few musical performances.

I followed my guidelines and built the songs one at a time, with the exception of one song, "You Can't Plagiarize Yourself," which fit the theme, but I wrote it 12 years earlier. Though I planned to write the album in a year, later songs took more time. The challenge to make each song more complex and interesting slowed my completion of the final three songs for almost another year. In the time that I worked through the end of the album, I joined a new band, Band of Waxx, whose players agreed to learn the songs so that I could perform them live. This resulted in two performances, a "practice show" on my 49th birthday and the "debut" performance of the first eight songs in January 2019 at the University Scholar Association connected to Pacific Lutheran University. I was supposed to perform the songs again solo at APS 2019 but lost my voice during the trip and could not sing.

Along the way, a student (research assistant, teaching apprentice, PLU Psi Chi vice president, coauthor) found the project so compelling that she decided to learn the songs as a way to learn piano too. Across the fall 2019 term, she and I performed the songs for my P242: Advanced Statistics and Methods students as they were intended. When a new faculty member joined the department in fall 2019 who knew bass, we decided to form the Purrfect Second Stringers (www.youtube.com/channel/UCov44ebsQcgS58MBCNR69Wg). The PLU Psi Chi chapter hosted the band to perform the entire album after the psychology department's fall research conference, which can be watched on video (www.youtube.com/watch?v=dbF-aPWkTzw&t=184s). Plans to perform again at WPA 2020 were interrupted by the pandemic. The ongoing pandemic interrupted the ability of the band to work together and stifled our ability to record. However, the live performances of the full band, also with the acoustic classroom performance, offer a glimpse into the fun these songs encourage while offering lyrics that help clarify fundamental concepts in the open science movement

CRISIS SCHMEISIS: SONGS TO INSPIRE SCIENTIFIC TRANSPARENCY

1. The Replication Crisis
2. Go Forth and Replicate
3. Preregistered
4. Decision Heavyweights
5. Ode to p-Hacking
6. You Can't Plagiarize Yourself
7. Becoming a Second Stringer
8. Open Science Alphabet
9. Progress: Open Science Promotes Diverse, Just, and Sustainable Outcomes
10. Scientific Transparency

Please use the following link to access the songs: https://osf.io/y2hjc/

1 A REPLICATION CRISIS

Responses Benefit Personal Workflow

Chapter 1 Objectives

- Define Replication Crisis
- Introduce causes of the Replication Crisis
- Conceptualize diverse, just, sustainable lens for science
- Explain why national parks are useful contextual examples
- Describe the open science movement
- Introduce tools and topics described later
- Introduce the book research question

MUSIC IN A BOX: ABOUT THE SONG "REPLICATION CRISIS"

The first song contains background information about the Replication Crisis, a series of events leading to major questions about the reproducibility of scientific findings. The lyrics introduce the setting of the album and consider some of the issues and problems that led to the crisis, ending with a nod to some early responses to the situation. This is the only song that names specific people and cheers them on for their part in initiating some changes to increase scientific transparency. The song offers a good list of scientists who pioneered open science for anyone who wants to look up their work. This song sounds like classic rock, but the minor key reminds the listener of the conflict of the crisis.

WHAT WAS THE REPLICATION CRISIS?

The beginning of this research methods journey, which aims to achieve scientific transparency in our work, started for many at the beginning of the Replication Crisis or "crisis of confidence" that emerged in the 2010s. Because many have been traveling this path for more than a decade, there are multiple retellings of the causes and consequences of this crisis (see Shrout & Rodgers, 2018). This book is personal in nature; the story is shared from my own experience within the crisis. This limited scope will result in a briefer description but does not intend to prioritize my singular narrative. Rather, the hope is that readers will face these questions from their own perspective and that this narrative will entice that interest.

In short, the Replication Crisis reflected concerns that published findings in peer-reviewed journals could not be replicated. Think on that problem for a moment. Textbooks, mental health treatments, educational interventions, and even public policies are drawn from research that is published using peer review. If the published findings cannot be trusted, then all the conclusions are suspect. To learn about the many causes, some of which will be explored in more detail later in the book, refer to the series of Special Sections in *Perspective on Psychological Science* (v. 7, #6, November 2012; v. 8, #4, July 2013; v. 9, #1, January 2014; v. 9, #3, May 2014), in which they are deeply explored. The first, entitled "Replicability in Psychological Science: A Crisis of Confidence," introduces the problem (Pashler & Wagenmakers, 2012; Pashler & Harris, 2012), potential explanations for why the problem existed (Makel, Plucker, & Hegarty, 2012; Bakker, van Dijk, & Wicherts, 2012; Ferguson & Heene, 2012; Giner-Sorolla, 2012; Klein et al., 2012; Neuroskeptic, 2012; Ioannidis, 2012), and recommendations for solutions (Frank & Saxe, 2012; Grahe et al., 2012; Koole & Lakens, 2012; Nosek, Spies, & Motyl, 2012; Wagenmakers, Wetzels, Borsboom, van der Maas, & Kievit, 2012). Across these manuscripts, one might draw a short list of causes as follows: (a) publication bias favors novel and unusual findings over replication and confirmatory research, (b) the presence of reward structures that favor many previous publications, and (c) poor reporting standards. Each of these is itself complex, with multifaceted causes, but the outcome is that research reports with flashy findings receive the greatest attention from both readers and researchers. The problem is that striving for those findings led to particularly inadequate practices in science.

POTENTIAL CAUSES OF THE CRISIS

These bad practices are highlighted in major events that occurred in 2011. Researchers often refer to 2011 as "the year of the crisis." Before these events, there was little concern for these problems in psychology, though some were voicing alarms more generally (Ioannidis, 2005). I myself had been pushing for reform for two years before these events, but no one really cared. After the year of crisis, I finally had an audience who was willing to help with "Harnessing the Undiscovered Resource of Student Research Projects" (see Grahe et al., 2012). Here is a brief description of two events that illuminated the replication crisis.

The most egregious affront against psychological science that alarmed the field in 2011 was when Diedrick Stapel was found to have falsified data in more than 40 published papers (Stroebe, Postmes, & Spears, 2012). This researcher was extremely influential, and his work is cited in many papers and textbooks. Over time, the lure of publication overwhelmed his ethics, and he started writing results sections with imaginary numbers. The papers were well written and interesting, but the findings were fiction.

Certainly, this man is not the only one who made up data or committed other forms of academic dishonesty. More critically, this example highlights a few problems with scientific reporting that need fixing. First, science reporting is built on trust. When manuscripts are submitted for peer review, reviewers are tasked with challenging the authors' rationale and methodology. They are expected to review and consider the results, but they are not expected to rerun analyses or review the quality of the data. While a reviewer might disagree with an author, authors' intentions are rarely questioned. This event highlights that in some circumstances, bad data and conclusions are due to willful disregard for scientific ethics.

However, another crisis event illuminates how bad science can emerge from good intentions. Daryl Bem published a paper in 2010 purporting to demonstrate precognition (parapsychological activity). Though there are many papers reporting the existence of parapsychological activity, this paper was published in the *Journal of Personality and Social Psychology*, one of the most prominent journals in social psychology. Further, Daryl Bem is a prominent social psychologist who suggested credible findings. Readers who believe in ghosts, goblins, astrology, tarot cards, and mind reading might be surprised to learn that this publication led to an uproar. Researchers demanded to see the data as they began to highlight many reporting issues evident in the manuscript. To his credit, Bem shared the data and did not argue strongly with the criticisms.

This second event introduces a number of related publication bias problems. Besides having a topic that is sensational and a prominent author that editors might favorably publish, the research was not maliciously reported. Bem did not intend to mislead or lie. Instead, his error was that he engaged in a series of questionable research practices more commonly described as hypothesizing after the results are known HARKing (Kerr, 1998) and p-hacking (Simmons, Nelson, & Simonsohn, 2011).

More critically, Bem was one of the researchers that taught the field how to effectively use these practices. In a book chapter about publishing an empirical article, Bem (2000) explains to future authors that, "There are two possible articles you can write: (a) the article you planned to write when you designed the study or (b) the article that makes the most sense now that you have seen the results" (p. 4). Bem argues that the correct answer is (b). Among otherwise good writing advice, Bem posits that the author should not bother a reader with the many pitfalls of the research practice. He suggests that rather than keeping a failed hypothesis in an introduction after conducting analyses, authors should rewrite a manuscript with new hypotheses and background literature to justify the findings that did emerge in the data. This is the definition of HARKing, but Bem argued that it was preferable to present a clear and straightforward story rather than distract the reader with errors made by the researcher. Later, in Chapters 3 and 5, this topic of massaging data to find effects, or p-hacking, and how to avoid it will be explored in more detail. For now, these events highlight that the challenges facing science were complex, while others would demonstrate that questionable research practices were both pervasive and systematic (Simmons et al., 2011; Bakker & Wicherts, 2011).

To understand these events, it is useful to remember that tools to make science easy to share are fairly recent. At the beginning of the new millennium, scientific manuscripts were still being submitted as hard copies, and journals published all materials in print, as there were no online journals or supplemental materials. With the cost of mailing documents and publishing printing pages, asking authors to also share data and materials was prohibitively expensive. Further, the drive toward shorter reports and, consequently, less stringent reporting standards was made in part to offer more publication opportunities for more authors as well as help disseminate findings and effects more broadly.

Regardless of the causes, this is a good moment to remind the reader that though this was publicly noted in social psychology, and many of the solutions were tested in social psychological research, the problem of publication bias and poor replicability pervades all fields of science, as suggested by Ioannidis (2005), who estimated that 50% of all published findings are false. In the decade that followed, many others recognized the need to change our approach to science, both in other psychological disciplines and also across the social and natural sciences.

WHY THE REPLICATION CRISIS DOES NOT MATTER

During 2017–2018, I completed a Crisis Schmeisis Tour to Increase Scientific Transparency. In almost 50 speaking engagements and meetings, I began my persuasive arguments with the position that it does not matter if there is a replication crisis in the field. Finding errors in methodology and improving them is the purpose of the scientific method. A good scientist avoids believing any truth, because the basic assumption is that our knowledge is only the best representation of truth, not the actual truth. From that perspective, one would expect publication errors, and our job is not to debate why they exist but, rather, how to do better science. This debate yielded tools and calls for change that will improve science and benefit the researcher at the same time.

While others continue to debate what effects may or may not be generalizable or whether replication efforts are appropriate or sufficient, my position has been and continues to be that there is greater benefit to learning new ways to be more transparent than there is in debating. Future scientific efforts will demonstrate what effects are generalized, but only if we move forward. This book focuses on this goal by introducing the reader to new tools and methods to conduct more transparent science. These tools include (a) new, free computer programs and software that make it easy to share plans, materials, and data; (b) research opportunities that collate resources and researchers to conduct more powerful research; and (c) reward structures that offer different paths to success. These tools are introduced through lyrics intended for both amusement and deep learning. Where possible, the examples consider the context of diversity, social justice, and sustainability while considering national parks. The objective of the examples is to connect the research methods content to ongoing social struggles with meaningful impact to the reader.

DIVERSITY, SOCIAL JUSTICE, SUSTAINABILITY, AND SCIENCE

Recently, I considered psychological research from the diversity, social justice, sustainability lens (Grahe, 2019). In that editorial, I used the metaphor of the "ideal" US national identity to describe and characterize how perspectives considering diversity, social justice, and sustainability differ from each other and yet need to be considered in whole when conducting and advancing science. By comparing the "US as a melting pot," "US as a salad bowl," and "US as a stewpot" approaches to the ideal integration of diverse cultures, I showed how diversity represents the various approaches and responses to those approaches, social justice represents who gets to determine policies that encourage these identities and how they are applied to people, and sustainability is the consideration of human and material resources in developing a nation and identity. I also presented a critical analysis of three recent special issues that advance a diversity, social justice, and sustainability lens in psychology.

A DIVERSITY, SOCIAL JUSTICE, AND SUSTAINABILITY LENS ON NATIONAL PARKS

Because this book uses national parks as an example, I reconsider that metaphor with the US National Park System. In this case, diversity represents both the distinct types of parks and the distinct peoples and uses they have for the parks. When examining types of parks, consider that the 419 properties in the US National Park System range from wonders of nature to historical locations. They are located across the country in rural and urban areas alike. To visit the parks is to understand the diverse geography and topography of the country. The diversity of locations and themes is dwarfed by the diversity of people visiting them and their experiences. Many parks are desired vacation destinations for global citizens, not just US citizens. Yellowstone National Park, the first national park anywhere in the world, boasts millions of visitors per year (Stats Report Viewer, n.d.). One study identified that almost one-fifth of the visitors were traveling from abroad (Warthin, 2017). As kids, on travel trips, we would try to find a license plate from each state before the trip was complete. When I played this game again with my own children, the pattern continued that most of our successes finding rare plates were in parking lots of national parks. The range of people brings a similar range of activities, as people have distinct ideals about what the perfect vacation is.

While diversity represents the breadth of unique experiences perceiving and engaging the world around us, social justice measures the degree to which all members have equitable access to the production and use of resources and experiences. When considering national parks, social justice topics include how the park was formed, the history of the park and what it represents, and who has access to current and future resources. In the United States, so many national parks that mark locations of wonder simultaneously represent great injustice, because the land was stolen from indigenous peoples through inequitable treaties that were often broken. For instance, local people were intended to be living inside Yellowstone National Park during its inception. As their presence unnerved visitors, the "indigenous" peoples were banned (Cunha & De Almeida, 2000).

Other concerns about social justice are modern, as rules about access and use are determined and implemented by government officials. Across time, the types of use within parks have changed as researchers identify human impact on wildlife and fauna. For instance, some rules that modern visitors experience include being forbidden from leaving the path in certain high-altitude locations to protect or restore alpine wildflowers, being required to camp in certain backcountry locations with severe limits on the use of fire and the disposal of waste, and not being allowed to take specimens of flora, fauna, or mineral. All of these restrictions are intended to increase park sustainability but represent differential justice across time. This highlights that justice limitations are complex, because sometimes limiting resource allocation for some visitors is intended to improve accessibility for other future visitors.

Diversity and social justice are easily considered in terms of human experiences, but valuing either requires sustainable approaches to extend both human and natural resources. The three rules mentioned about limiting modern use were enabled to increase sustainability. In each case, these were about extending natural resources. But human resources also need to be managed sustainably. There need to be enough park rangers and staff to run a park, and they need to be adequately resourced. For example, when there is insufficient labor to complete all necessary work, bridges crossing rivers and streams are not immediately replaced when destroyed or damaged, and trees blocking trails are not removed. Thus, in parks, sustainability references not only the flora, fauna, and geography but also the people managing, living near, and using the parks. Understanding by thinking about the diverse world of national parks helps clarify the concept, but Grahe, Cuccolo, Leighton, and Cramblet Alvarez (2019) explain not only how each is important in psychology but also how using open science initiatives can improve their representation in the field. This is discussed more fully in Chapter 9. Presently, it is sufficient to begin using a diversity, justice, and sustainability lens to consider the topics and initiatives described in this book.

CRISIS SCHMEISIS BOOK REVIEWS

When completing the Crisis Schmeisis tour, being on sabbatical allowed me to read books that I normally would not have time to read. I chose books that were historic or narrative rather than psychological to help me think about problems in different ways. I mostly found these books at national parks' bookstores by selecting topics that would compel me to think about diversity, justice, and sustainability. While most of the books were focused on indigenous peoples, racial justice, or women, in this first chapter, I offer a book review about a project that was intended as a replication project itself.

Chapter 1: Crisis Schmeisis Journey Book Review

The Man Who Planted Trees: A Story of Lost Groves

Jim Robbins

This book is about a man who decided he wanted to replicate old-growth trees. Read my book review and decide if you want to explore further (https://osf.io/tz6e7/).

WHAT IS OPEN SCIENCE?

If this book is a diversity, justice, and sustainability–inspired musical journey through open science and anchored in national park examples, it is important to explain what aspects of science researchers and practitioners hope to make more transparent. Readers of this book will likely be familiar with the basic steps of the scientific method. Inherent in these steps is an assumption that others can then repeat the measurement and analyses to draw the same conclusions. When those conclusions differ, researchers seek to explain those differences through further testing and experimentation.

One goal of this book is to help inform the reader so that they can successfully conduct research transparently. When considering the scientific process, all steps should be transparent so that others can repeat and further test the research that is conducted by anyone else. As the book progresses, I use the terms "open science" and "scientific transparency" interchangeably. These terms are defined by the process of conducting science following procedures that yield reproducible research. The Transparency and Openness Promotion Guidelines (Nosek et al., 2015) were written by researchers, publishers, and granting agencies with the aim of helping others become more aware of transparency of scientific research.

The open science movement is a partial response to the replication crisis. While some were advocating to increase openness before, most of the change has occurred since 2012. The movement is trying to change science in many ways, but they can be grouped into three categories: (a) open access to make knowledge available to all, (b) transparency tools to advance reproducibility, and (c) collaborative processes to increase efficacy and access. This book will describe in detail the transparency tools and collaborative processes. Given that the reader was asked to pay a small fee to access this material, I demonstrate that my bias for transparency does not extend to open access. In short,

Table 1.1 Scientific Method

I.	State a problem
II.	Study past research
III.	State hypotheses
IV.	Measure
V.	Analyze/test hypotheses
VI.	Draw conclusions
VII.	Report findings
VIII.	Repeat

there is much work required in creating scientific reports and other scientific knowledge. While I support mechanisms that make research broadly available, beyond this section, the reader will see little more about open access from me.

Historically, only some impediments to openness included technical barriers associated with research before the digital age. Further, researchers themselves impede transparency when they choose to not share their research out of fear, jealousy, or some other human trait. Another set of obstacles to openness is related to the monetization of knowledge when individuals, companies, and other organizations own portions of the process, whether that be the measurement tools, the data, the analysis software, or the publishing outlets.

OPEN SCIENCE INITIATIVES IN FOCUS

With the broad goals of open science clarified, it is important to recognize that the scope of this book is limited due to length and the focus on benefitting beginning scientists. Therefore, the reader will return to the following initiatives across multiple contexts: (a) project management software, (b) planning tools and expectations, (c) data and materials transparency, (d) changes in results reporting standards, and (e) crowdsourcing research. Readers will find a brief description of each here at the close of this chapter. For my own work, I have adopted the Open Science Framework (OSF) for my project management. In

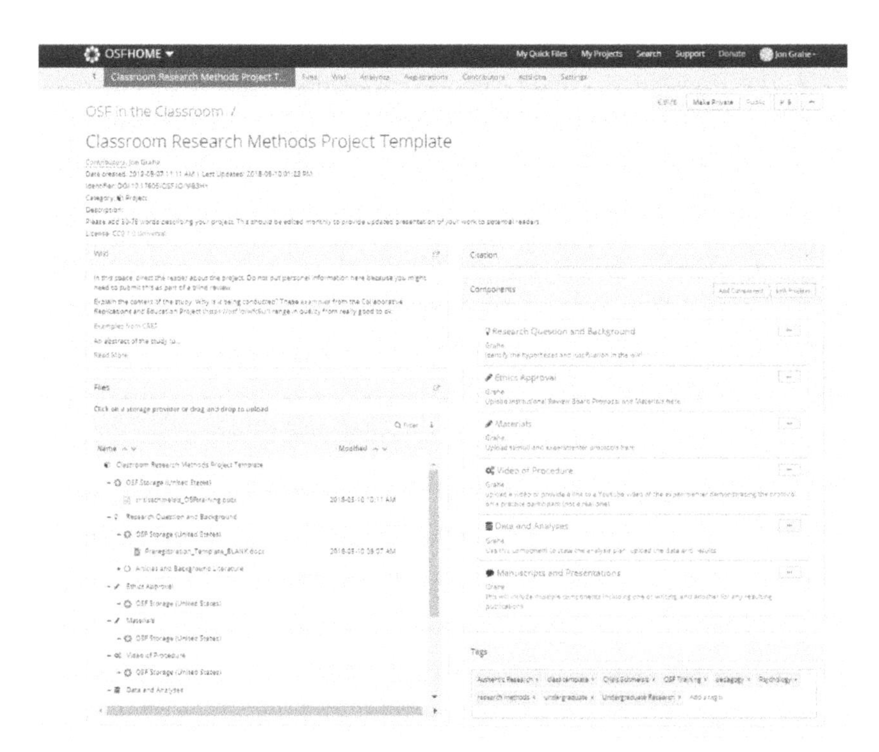

Figure 1.1 OSF Page: Classroom Research Methods Project Template

fact, I encourage students to use my Classroom Research Methods Project Template to build their own projects if they do not have one already started (https://osf.io/mb3hy/). Exercise 1.1 provides some instructions on using this template.

The OSF is a software platform developed by the Center for Open Science to help make research transparency easy to integrate into a researcher's workflow. Changes in the way we plan research include the concept of preregistration, which requires us to prepare much more of a project before data are collected than in traditional methods. Along with this preplanning in the form of preregistration is the expectation that researchers will share both their materials and data that accompany a research report. Open Science Badges are incentives to alert readers that an author is sharing research plans, data, materials, or all of the above. Changes in reporting standards require more than just F and p-values to convince readers of strong inference. Effect sizes, confidence intervals, and meta-analyses offer an alternative method of examining data and results that are less biasing than p-values. When looking at others' data, one can also conduct a p-curve analysis to determine potential publication bias in a set of published findings by looking at the distribution of probability values. Finally, crowdsourcing research projects offers an opportunity to collate the work of many researchers to study one problem (or set of problems).

RESEARCH IN ACTION: INTRODUCING THE BOOK RESEARCH QUESTION

This book is organized to help a beginning researcher prepare, organize, and execute transparent research. While each chapter will include general descriptions of how to conduct this research, there will also be a recurring example that follows the steps of the book. And to stay consistent with the book's theme, the ongoing research question will be, "What characteristics predict the number of miles hiked by visitors at national parks?" This project could provide applicable data, and there are several methods for answering this question. This allows for many different research questions to be advanced by readers at different stages of the educational process. For instance, an introductory study might approach this with an observational design with limited variables and simple analyses. A more advanced student might design an experiment and try to change behavior through direct manipulation. Still more advanced students might mix survey and experimental methods to describe a complex model with many predictors.

This research question is meaningful beyond a simple learning example. Knowing how people use the parks, particularly trails located far from permanent structures, could allow the National Park System to make decisions about where and what resources should be focused in any particular area. In future chapters, I will include exercises to help move a project forward and to experience open science tools, including, among others: (a) presenting a research question, (b) conducting a literature review, (c) hypothesis generation, (d) research design, (e) a p-curve analysis of related research, and (f) documenting the project on the OSF.

CHAPTER 1 EXERCISES

Exercise 1.1a. Identifying and Preparing Your Question for Novel Research

- Make a list of psychology courses you have taken.
  - Review and list the most interesting topics and pick one.
  - Read one or more primary research articles related to that topic.
- Think of three research questions that are significant to you.
- Talk about research ideas with peers and friends till refined.
- Choose one of those questions, and find and read 5 to 10 articles on the topic.
  - Read reference sections to find related articles.
  - Search Google Scholar for manuscripts that cited related articles.
- Consider the evidence presented in the articles and generate research hypotheses.
  - Determine what methods you would use to answer your question.
  - Define your variables, describe your procedure, and then determine how you will run your analysis.

Exercise 1.1b. Identifying and Preparing Your Question for a Crowd Research

- Review list of Collaborative Replications and Education Project, Network for International Collaborative Exchange, StudySwap, or alternative projects offered by the instructor.
  - Pick one that most closely aligns personal interest with professional goals while considering feasibility.
  - Read the original or target article associated with the effect being studied.
  - Read the Method section again to become extra familiar with protocols.
- Review the reference list of the target article and identify potential supporting research.
- Search for the article in Google Scholar and review list of manuscripts that cited target article.

Exercise 1.2. Make an OSF Project for Research Project

Use the OSF: For individuals who want a very basic tutorial on OSF, use (http://osf.io/p4fse/), the page I built for authors at the *Journal of Social Psychology* who needed help. Also, the Center for Open Science has a YouTube channel with many valuable videos, including "Getting Started with the OSF."

Follow these instructions: to begin your OSF project.

- Fork or duplicate OSF pages from my research methods template (https://osf.io/mb3hy/).
  - Forking makes a new copy that is tethered to the original.
  - Duplicating copies the page and all its contents, but it is not tethered to the original.
  - This page includes instructions for each component to follow through the research process.
- Follow the template instructions to make the page your own, and edit the OSF page to include the following:
  - Add a title: click on the title and make a short name for your page.
  - Add a description: below the title, use this description section to write a short abstract (<50 words).
- Edit the wiki.
  - In the OSF, the wiki allows the researcher to communicate with the reader. Edit your wiki on the main OSF page to communicate basic information about the project.
- Add the following:
  - Restate your title.
  - State your research question.
  - Add a 120- to 150-word abstract.

Example 1.3: Review the Book Project Example OSF Page

- Visit the Journey into Open Science Book Project Example (https://osf.io/b4nrh/).
- Review the changes made from the template.
- Critically evaluate the presentation and content presented on the page.
- Identify things to do differently or keep the same on you own project.

END-OF-CHAPTER REVIEW QUESTIONS

1. Compare and contrast the major events that brought the Replication Crisis to light.
2. Name three approaches to ideal integration of diverse cultures. Which makes the most sense to you?
3. List and describe the components of diversity, justice, and sustainability.
4. What piece of common publishing advice often led to questionable research practice?
5. What other possible problems, besides the ones listed in this chapter, could result from the advice that Bem gave about writing?

6. Do you think that there were other obstacles to sharing and communicating scientific understanding besides the ones mentioned in this chapter? How would those obstacles contribute to the severity of the crisis?
7. What are ways open science connects to the principles of diversity, justice, and sustainability?
8. What are some other potential factors that could have led to the crisis of confidence in scientific findings?

References

Bakker, M., van Dijk, A., & Wicherts, J. M. (2012). The rules of the game called psychological science. *Perspectives on Psychological Science, 7*(6), 543–554.

Bakker, M., & Wicherts, J. M. (2011). The (mis) reporting of statistical results in psychology journals. *Behavior Research Methods, 43*(3), 666–678.

Bem, D. J. (2000). Writing an empirical article. In R. J. Sternberg (Ed.), *Guide to publishing in psychology journals* (pp. 3–16). Cambridge: Cambridge University Press.

Cunha, M. C. D., & De Almeida, M. W. (2000). Indigenous people, traditional people, and conservation in the Amazon. *Daedalus, 129*(2), 315–338.

Ferguson, C. J., & Heene, M. (2012). A vast graveyard of undead theories: Publication bias and psychological science's aversion to the null. *Perspectives on Psychological Science, 7*(6), 555–561.

Frank, M. C., & Saxe, R. (2012). Teaching replication. *Perspectives on Psychological Science, 7*(6), 600–604.

Giner-Sorolla, R. (2012). Science or art? How aesthetic standards grease the way through the publication bottleneck but undermine science. *Perspectives on Psychological Science, 7*(6), 562–571.

Grahe, J. E. (2019). Checking progress toward a more diverse, just, and sustainable psychology. *Psi Chi Journal of Psychological Research, 24*(2), 77–83.

Grahe, J. E., Cuccolo, K., Leighton, D. C., & Cramblet Alvarez, L. D. (2019). Open science promotes diverse, just, and sustainable research and educational outcomes. *Psychology Learning & Teaching, 19*(1), 5–20.

Grahe, J. E., Reifman, A., Hermann, A. D., Walker, M., Oleson, K. C., Nario-Redmond, M., & Wiebe, R. P. (2012). Harnessing the undiscovered resource of student research projects. *Perspectives on Psychological Science, 7*(6), 605–607.

Ioannidis, J. P. (2005). Why most published research findings are false. *PLoS Medicine*, 2(8), e124.

Ioannidis, J. P. (2012). Why science is not necessarily self-correcting. *Perspectives on Psychological Science, 7*(6), 645–654.

Kerr, N. L. (1998). HARKing: Hypothesizing after the results are known. *Personality and Social Psychology Review*, 2(3), 196–217.

Klein, O., Doyen, S., Leys, C., Magalhães de Saldanha da Gama, P. A., Miller, S., Questienne, L., & Cleeremans, A. (2012). Low hopes, high expectations: Expectancy effects and the replicability of behavioral experiments. *Perspectives on Psychological Science, 7*(6), 572–584.

Koole, S. L., & Lakens, D. (2012). Rewarding replications: A sure and simple way to improve psychological science. *Perspectives on Psychological Science, 7*(6), 608–614.

Makel, M. C., Plucker, J. A., & Hegarty, B. (2012). Replications in psychology research: How often do they really occur? *Perspectives on Psychological Science, 7*(6), 537–542.

Neuroskeptic. (2012). The nine circles of scientific hell. *Perspectives on Psychological Science, 7*(6), 643–644.

Nosek, B. A., Alter, G., Banks, G. C., Borsboom, D., Bowman, S. D., Breckler, S. J., Buck, S., . . . Yarkoni, T. (2015). Promoting an open research culture. *Science, 348*, 1422–1425.

Nosek, B. A., Spies, J. R., & Motyl, M. (2012). Scientific utopia: II. Restructuring incentives and practices to promote truth over publishability. *Perspectives on Psychological Science, 7*(6), 615–631.

Pashler, H., & Harris, C. R. (2012). Is the replicability crisis overblown? Three arguments examined. *Perspectives on Psychological Science, 7*(6), 531–536.

Pashler, H., & Wagenmakers, E. J. (2012). Editors' introduction to the special section on replicability in psychological science: A crisis of confidence? *Perspectives on Psychological Science, 7*(6), 528–530.

Shrout, P. E., & Rodgers, J. L. (2018). Psychology, science, and knowledge construction: Broadening perspectives from the replication crisis. *Annual Review of Psychology, 69*, 487–510.

Simmons, J. P., Nelson, L. D., & Simonsohn, U. (2011). False-positive psychology: Undisclosed flexibility in data collection and analysis allows presenting anything as significant. *Psychological Science, 22*(11), 1359–1366.

Stats Report Viewer. (n.d.). Retrieved February 1, 2021, from https://irma.nps.gov/STATS/SSRSReports/Park%20Specific%20Reports/Annual%20Park%20Recreation%20Visitation%20(1904%20-%20Last%20Calendar%20Year)?Park=YELL

Stroebe, W., Postmes, T., & Spears, R. (2012). Scientific misconduct and the myth of self-correction in science. *Perspectives on Psychological Science, 7*(6), 670–688.

Wagenmakers, E. J., Wetzels, R., Borsboom, D., van der Maas, H. L., & Kievit, R. A. (2012). An agenda for purely confirmatory research. *Perspectives on Psychological Science, 7*(6), 632–638.

Warthin, M. (2017). *Yellowstone releases reports about visitors and traffic*. Retrieved from www.nps.gov/yell/learn/news/17042.htm

2 GO FORTH AND REPLICATE

Making Methods for Others

Chapter 2 Objectives

- Define replications
- Examine benefits and problems
- Discuss types of replication
- Define success with a replication
- Consider replications from national park examples
- Describe large-scale metascience replications
- Examine collaborative replications in light of diversity, justice, and sustainability

MUSIC IN A BOX: ABOUT THE SONG "GO FORTH AND REPLICATE"

The second song of the album, "Go Forth and Replicate," introduces the importance of replication in science and how to conduct large-scale crowd projects. One solution to the crisis is more replication, and crowdsourcing replications became more frequent as a way to reexamine effects with more diverse researchers and samples. This song invites everyone to engage in replication and provides a basic framework for organizing a large-scale replication. It even highlights some unfortunate outcomes to joining replication, such as level of authorship, but the focus is on the benefits of generalizable conclusions. This song is inspired by gospel and punk and calls the audience to join the movement.

WHAT REPLICATIONS ARE AND ARE NOT

Rarely is it appropriate to use a common dictionary to understand a discipline-specific topic, but in the case of replication, it is a great beginning. The common definition of replication is a noun reflecting an action of copying or reproducing something. There are definitions related to law and genetics, but for most of us, the word "replication" has the same meaning. In research methods, this definition also applies. We will also use the word as a verb when a researcher is replicating someone else's work. Finally, the product of that replication process, the research study, is often referenced as an object called a replication. That would make the following sentence correctly written, if awkward to read: "Taking part in the **replication process**, Grahe et al. (2020) **replicated** Fosse and Toyokawa (2016), and readers can access materials and data from that **replication** on the EAMMi2 project page (http://osf.io/te54b/)." Except as an example of these three meanings, it is bad form to include the same word repeatedly in a sentence. However, the reader now recognizes our shared understanding of the term "replication." In science, replication means trying to reproduce other researchers' work.

Whereas the basic definition of replication remains the same, the science of replication has been brought to many scholars' attention due to reproducibility and replicability of scientific research. Right at the beginning of the Scientific Revolution 2.0 (Spellman, 2015), there were arguments about whether replication attempts were good enough to represent the original effect (Maxwell, Lau, & Howard, 2015; Anderson & Maxwell, 2016) and whether the effect sizes represented successful replications of the original effect (Patil, Peng, & Leek, 2016). Finally, if one replication is good, many replications are better, and a major outcome of the open science movement has been large-scale collaborative replications or crowdsourcing research projects (Uhlmann et al., 2019). This chapter will introduce each of these in more detail. Aspiring researchers should anticipate that this complex topic is not quickly mastered, as data scientists continue to refine and interpret replications.

REPLICATION BENEFITS SCIENCE

While the titles for each step of the scientific method, or even the exact number of them, might vary slightly from text to text or discipline to discipline, the basic steps remain the same (state a problem, study past research, state hypotheses, collect data with appropriate measures, analyze/test hypotheses, report findings, repeat). It is that last step that is the focus of this chapter: repeat. Replication is the process of repeating the science that others have done. We might build on that theory with alterations to the materials or procedures, but there is an assumption in science that research reported should be generalizable, repeatable, replicable, and reproducible so that scientists can improve rigor and transparency in scientific research.

Chapter 1 introduced the replication crisis as a time when we started questioning the reproducibility of published work. Though much of this is associated with social psychology in particular, one of the most-read and -cited critiques of research findings was from Ioannidis (2005), who trained in medical fields and spoke about science generally. As with all movements, there are those who argue that their own discipline is immune. While trying to convince others to adopt these procedures in their own work, peers in other subdisciplines denied their field had problems, as evidenced by some reason or another. Then as now, this debate is less compelling than whether to make our science more open and more applicable. So I return to the basic argument that the last step, repeat, requires them to show all their data, materials, and plans. I wonder how many will still be resisting in another 10 years.

PROBLEMS FROM SINGLE-STUDY PROJECTS

One of the causes of the replication crisis is that peer-review journals are competing for readers and authors. One marker of a journal's success is the one-year impact factor based on the average number of times articles were cited over the previous cycle. A bigger impact factor means more people used that article in other research. As the problem continued to manifest, journals were less interested in publishing long articles that included many studies (their own replications), and there had never been much interest in publishing replications of other people's work. Increasingly, word counts decreased, and the desire for flashier, counterintuitive research led to little interest in replication as a field. Furthermore, the work that was published did not include as much information as authors cut words to meet editors' demands. This has two clear problems. The first is that readers are left with relatively little information necessary to carry out a replication. The second is that the reviewers and editors were not seeing all the information about the research being published. Chapter 5 presents the problems of HARKing and p-hacking in more detail. For now, please just note that they are existing problems with single-study research papers.

The more pressing issue associated with relying on single-study papers to inform theory or guide policy is that any single estimate of an effect is decidedly insufficient. In statistics, the Law of Large Numbers states that the bigger the sample size, the closer the estimate is to the true population value, and the Central Limit Theorem states that the sampling distribution of sample means becomes more normal with increasing sample size. Taken together, they suggest that any study should have a large sample size. Sometimes this can be achieved with a single study. Projects with $N > 10,000$ from a single sample, or better yet, $N > 1,000,000$, offer compelling estimates of the general population. However, contrast that with the common instruction for experiments to have sample sizes of at least 10 or 20 per condition. These numbers are woefully insufficient to draw strong inferences about the larger population. Instead, it is better to assume that the effect would be most applicable to that local population until demonstrated again. If demonstrated in the same local population, the average effect size is a better estimate of that group. If demonstrated in other locations, the average effect size is more generalizable across groups. If the purpose of research is to explain 7+ billion human beings, samples with 40, 80, or even 160 subjects from one location are hardly sufficient. By replicating more studies, science is taking advantage of statistical principles that demonstrate better outcomes with more subjects and more trials.

TYPES OF REPLICATION

There are different approaches to replicating someone else's work. The first approach is to follow the same protocols, use the same materials, and match the original study in every way possible. This idea of an **exact** replication directly contrasts with a **conceptual** replication approach, which tests the same theory but makes some changes to the materials or protocols to further advance theory in some way. This basic dichotomous categorization does not fully describe these two types, however. One important consideration is that conceptual replications cannot test an original effect, because any difference in outcomes can be attributed to confounding factors between the original study and the conceptual replication. But the major problem is that exact replications are hard to achieve. Without time travel, a replicator cannot access the same population. Even if time confounds were ignored and the differences in location and time of testing were discounted, differences in researchers, methods to display stimuli, and approaches to randomization compound the unique status of the replication. All these differences cannot be ignored in sum, so the idea of exact replications must be addressed by trying to achieve conditions that match wherever possible and recognizing the differences as part of the reporting process. These **direct** or **close** replications provide the strongest test of an original effect. However, they are not conclusive by themselves (see Figure 2.1 to consider different replications).

A highly powered direct replication that is able to detect very small effects is only one more attempt at discovering the true effect size that is being studied. Conclusive evidence follows from many replications conducted by many researchers with many populations. It is only through aggregating these many effects and evaluating them through meta-analysis that we can make confident conclusions about the nature of an effect. The topics of effect sizes and meta-analyses are discussed more later, so it is only necessary to recognize that any strong inference about a theory needs many tests of the effect, not just one or a few.

Further, for these meta-analyses to yield unbiased conclusions, the analysts need to know as much about the data from each replication as possible. Documenting each condition of the study and its differences from the original can be achieved through the use of a checklist offered by the authors of "The Replication Recipe" (Brandta et al., 2014). Soon after the crisis, this group of researchers identified five critical components to conducting a successful replication: (1) Carefully defining the effects and methods that the researcher intends to replicate; (2) following as exactly as possible the methods of the original study (including participant recruitment, instructions, stimuli, measures, procedures, and analyses); (3) having high statistical power; (4) making complete details about the replication available so that interested experts can fully evaluate the replication attempt (or attempt another replication themselves); and (5) evaluating replication results and comparing them critically to the results of the original study.

To achieve these ingredients, the group offered 36 questions to guide decisions and excellent reporting in six areas (https://osf.io/2wznj/). Table 2.1 shows some of those questions, with the entire 36 questions available at the original publication (see Table 1, Brandta et al., 2014).

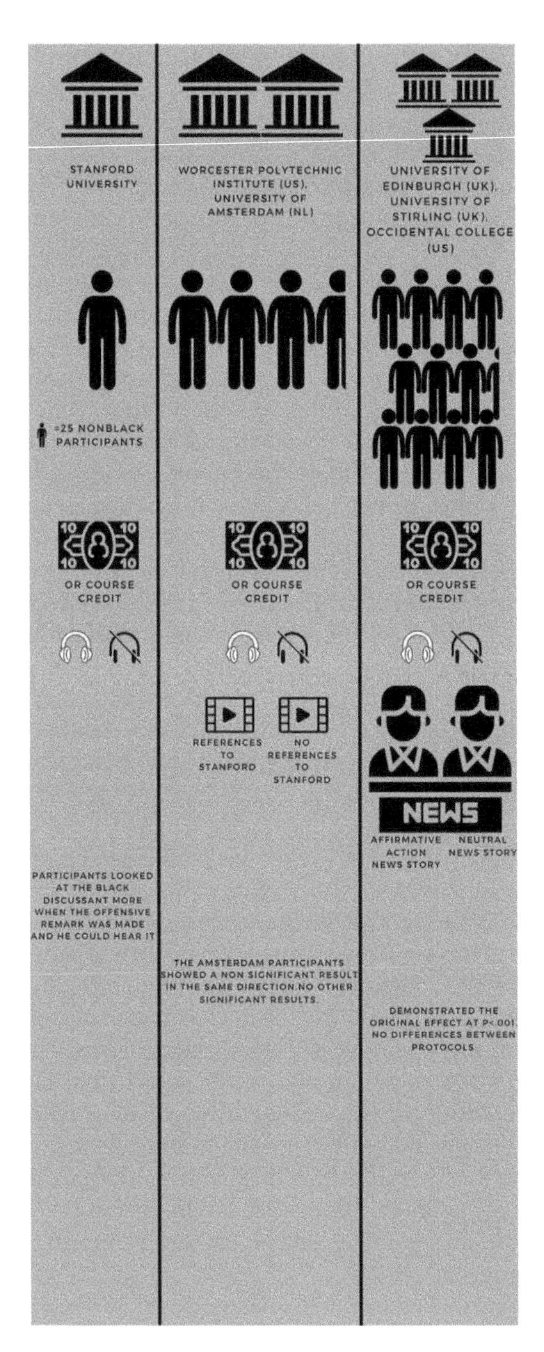

Figure 2.1 Types of Replications

Table 2.1 Example Questions from the Replication Recipe (Brandta et al., 2014)

The Nature of the Effect	
1. Verbal description of the effect I am trying to replicate:	
2. The effect size of the effect I am trying to replicate is:	
Designing the Replication Study	
12. Location of the experimenter during data collection:	
16. The rationale for my sample size is:	
Documenting Differences between the Original and Replication Study	
20. The similarities/differences in the procedure are:	[Exact I Close I Different]
23. The similarities/differences between participant populations are:	[Exact I Close I Different]
Analysis and Replication Evaluation	
26. My exclusion criteria are (e.g., handling outliers, removing participants from analysis):	
27. My analysis plan is (justify differences from the original):	
Registering the Replication Attempt	
29. The finalized materials, procedures, analysis plan, etc. of the replication are registered here:	
Reporting the Replication	
33. I judge the replication to be a(n) [success/informative failure to replicate/practical failure to replicate/inconclusive] (circle one) because:	
35. All of the analyses were reported in the report or are available here:	

DEFINING A SUCCESSFUL REPLICATION

Should a team of researchers convey to you that they successfully replicated an effect, what is the first thing that comes to mind? Or conversely, if they were unsuccessful, what does that imply to you? Does a successful replication represent a statistically significant finding and the converse a lack of rigor in science? The other valid interpretation is that the researchers are reporting that they correctly completed all the procedures using the appropriate materials. They are not speaking about the outcome at all but rather about the completion of the data collection. It could be that the results that emerge are not statistically significant at all or that they could be in the opposite direction from the original. Finally, it is possible that the replication was not statistically significant, but the effect was in the same direction and not different from the original. Successful replication means different things to different researchers, so readers should beware of bold statements and pay close attention to the outcomes themselves. As a scientific researcher, you can avoid this problem by stating that you successfully replicated the methods or the results to keep readers from getting confused.

PARKS IN ACTION: NATIONAL PARK SYSTEMS AS A METAPHOR FOR REPLICATION

Elsewhere, in order to help the research community recognize that Collaborative Replications and Education Project (CREP) rhymes with grape and not gap, I described replications using grapes as the metaphor after creating a logo of grapes (see Figure 2.2) by noting their differences in size, color, or taste on the same bunch. However, for this book, the more complex National Park System (NPS) provides a richer demonstration of how a group can aim to achieve similar outcomes to replicate while the subjects' experiences are variable and distinct. Before inviting you to imagine this metaphor, I offer this simple research hypothesis: "There is a curvilinear relationship between age and distance the average visitor walks when visiting a national park such that individuals in the middle age groups will walk more than those at either extreme." To test this question, we need to collect visitors' ages and the number of miles they walked, then compute the correlation between age squared with miles. If this correlation is positive and different than zero, that would suggest support for this hypothesis. To compare across national parks, a plot of 95% confidence intervals (CIs) will offer a clear display of potential differences. Plots of CIs

Figure 2.2 The CREP Logo

Figure Caption: Grapes are a great metaphor for replications because a single vine will include many bunches of grapes. Inspecting them will reveal differences in size, shape, and ripeness. Just like close replications, differences still emerge. Akin to conceptual replications, different species of grapes offer qualitatively different flavors and textures.

around effect sizes (such as d or r) are also consistent with how replications are presented in meta-analytic work (see Cumming, 2014; Cumming & Calin-Jagerman, 2016). From a scientific perspective, we want to know whether this effect is consistent across the population. To do that, we need to conduct replications. Imagine being that researcher, going to the nearest national park, and standing at the visitor center to administer the survey to randomly selected visitors. Throughout the book, this study will expand with many hypothetical considerations, and the reader is invited to further explore possibilities.

Link to NPS list of parks: www.nps.gov/aboutus/national-park-system.htm.

To help with this imaginary task, readers can understand the true breadth of the NPS by reading a quote from the FAQ page regarding the 419 areas they manage: "These areas include national parks, monuments, battlefields, military parks, historical parks, historic sites, lakeshores, seashores, recreation areas, scenic rivers and trails, and the White House" (www.nps.gov/aboutus/faqs.htm).

To limit the scope of the metaphor, this book only considers the national parks, those entities created by Congress. Their formation and their scope distinguishes them from the many National Historic Park or Site ($N = 133$), Battlefield Park or Site ($N = 25$), National Monument ($N = 84$), or other designations ($N = 116$). National parks are also easy to imagine as leisure destinations. Yellowstone represents the very first national park anywhere in the world. It was a concept realized by the creation of Yellowstone National Park. In the century that followed, 61 more were added.

As with the first national park, most exist to offer public access to natural wonders. The Gateway National Arch is a creation of architects, engineers, and construction workers built across the Mississippi River to reflect westward expansion. Another, Mesa Verde National Park, is a testament to architects, engineers, and construction workers of the Pueblo people 1,000 years before. Their homes carved in the sides of cliffs reflect small apartments in prime real estate near precious resources for a long-gone civilization. Even when considering that the bulk of the parks protect wonderous landscapes and geology, each and every one also reflects the humanity of the people who used these spaces for life-supporting or leisure activities for many past millennia. As such, it is important to reflect on that history when considering modern experiences of those spaces.

These people across time differed in so many ways, and so much of their histories were lost from the genocidal activities of the European settlers who arrived and the devastating effects of the diseases they brought with them. However, even those settlers differed in many ways from countries and cultures across the world and even had differing beliefs within the same culture. While the contributions of people of color are often ignored, immigrants from across the world were drawn to the dream of the New World. Not all people who contributed to the history of the parks did so willingly. The work of enslaved people can be seen in many directions from Shenandoah National Park, and their migration toward other national parks followed freedom. In short, these parks often convey stories of very diverse people from across many distinct times. From a generalist replication perspective, consider how the NPS employees are challenged with presenting information and conveying these varied stories.

For the present study, these details directly relate because they reflect the diverse histories across distinct national parks, and they reflect different goals of visitors, with some inviting more walking than others. If people are arriving from across the world to visit US national parks (more than 330,000,000 visitors in 2019), consider all the variables that might impact differences in their individual walking. However, for the present hypothesis (H1: a positive quadratic effect between age and number of miles walked on average by visitors), this subject variability will be ignored for the sake of simplicity. I predict this effect to be large enough to exceed any error variance caused by individual differences. However, because we expect large individual variability, the effect should be estimated to be small, because we are not controlling for all individual differences. Hence, if I believe that the relationship between age^2 and visitor miles to be the true effect with $r = .40$, I would expect it to be smaller, because I am studying it without controlling for any **covariates**. To be conservative, I will estimate an $r = .20$ for this effect size when I make the power estimate in Chapter 3.

Now, imagine this study again, and think about replication. For the formulas, assume that each park has a normal distribution of visitors, and then guess how many miles a visitor would walk on average, for each park (two columns: age and miles). While correlations are likely zero, that reflects random connections between the two formulas. Rather, the exercise helps the reader practice correlations and imagining the challenges of replicating a simple study across many locations. While the data are available, a student might also quickly calculate the *t*-tests, ANOVAs, and estimate effect sizes with 95% CIs for practice.

When considering the National Park System as a metaphor for direct replication, students should think about controlling how the researchers are dressed or how they speak to the respondents in a real-life situation. They also need to think about the location in the visitor's center and the methods of random selection. Other factors such as time of day, time of year, type of weather, other conditions in the park such as fire, and even the financial state of the world can also impact the scores. In order to test the hypothesis correctly, all these factors, in addition to some of those individual differences we have ignored to this point, all need to be controlled for either methodologically or statistically.

Finally, imagine doing this study alone. It would not be possible within an adequate time frame to get this data collected in all the parks in one summer alone. It would also be extraordinarily expensive to do it as a solo researcher. Thus, this example highlights my true passion with students, engaging in large-scale collaborative research using open invitation and open science research projects.

LARGE-SCALE REPLICATIONS

I hope that the readers of this book have an opportunity to contribute to one of the project sites mentioned. One of the best outcomes of the Replication Crisis is the emergence of large-scale collaborative research in psychology. While all systems need refining, these projects offer large benefits to the researchers that participate, as they address major questions for science. Further, the data they generate are normally publicly available for others to access and answer as-yet-unidentified questions. While these projects are impressive and remarkable to witness, they are also challenging to manage. However, with an operating budget of zero dollars since its inception, the CREP has become an impressive organization based solely on the voluntary contributions of the reviewers, instructors, and students who are part of it. Though the labor costs have been enormous, they are generally lined up with the goals and needs of the contributors and thus benefit their own educational and professional journey as they generate valuable effect size estimates for the purpose of science.

MANAGING CONTRIBUTORS THROUGH TIME AND SPACE

The reader of this book might not be working on contributing to one of these projects, and open science does not require collaboration (as one reviewer accurately reminded me on Grahe, Cuccolo, Leighton, & Cramblet Alvarez, 2019). However, that does not mean that researchers work alone. Most research labs include multiple collaborators even if there is one person in charge or only one person named on the manuscript. It is rare for a solo researcher to actually conduct every aspect of a research project with no one else involved, though it does happen. And yet this solo researcher still is not alone. Even this rare person needs to clearly communicate to the reader about all aspects of the research so that meta-analyses can appropriately consider the context. Further, a solo researcher, at the very least, needs to document the study details to communicate with the future self and collaborators.

Any researcher can easily forget small, minute details, e.g., which idea came from which paper, why subjects were deleted from the data set, the explicit reason that certain variables were chosen for the final dependent variable, and exact methods of transforming data and why. Even the best researcher is human, with biases and memory loss. Approaching all aspects of the research process as though we have collaborators that need to be able to explain it to someone else will save time if the manuscript is accepted for publication or headaches when work needs to be redone. Personally, I have found great success using a combination of hosting projects on the OSF while using Google Docs and Google Forms to collectively write materials and manuscripts or to disseminate and collect information. There are other approaches, but these are both free, with short learning curves. Thus, I will use them through this text as the default location for projects and work.

FINDING THE "RIGHT" PROJECT

Students need the entire term to complete a sample for some of these projects. Often, a complete sample can require students to continue data collection as an independent researcher when research pools are limited or the semester is too short for students to get all the necessary work done. Therefore, the earlier the selection process begins, the better. The most likely options for a student would be the CREP, the NICE, or the Psychological Science Accelerator (PSA; Moshontz et al., 2018). However, single-time projects such as the Many Labs (1, 2, 3, 4, or 5) or EAMMi2 might be announced at any time. Researchers need to find projects that fit their goals and resources.

The CREP offers six to eight different studies selected specifically for students to complete in an academic term, and students regularly succeed at these projects. The NICE offers projects each year, again with undergraduates in mind. Most other options (such as the PSA or Many Labs) are only feasible for advanced undergraduates who are working in a faculty lab after completing the initial research methods course. The CREP and NICE combination offers students many potential choices that vary in question type and methods. In all cases, contributors are offered step-by-step instructions to complete a sample worthy of the larger project.

The researcher needs to consider their own goals. Most psychology majors directly enter the workforce and not graduate school, and only a fraction of those who attend graduate school are focused on becoming research psychologists, so very few students are driven by the notion of authorship on a paper (particularly when they will be a minor author among dozens of co-authors). Instead, students can choose projects that will be used by other researchers for the major paper. If this level of contribution is not intrinsically motivating to the student (my own personal experience suggests that about half of my students get excited by the idea of making an impact by sharing their sample, and less than 10% seek any authorship opportunities), there are the added bonuses of learning to excel in data management and analyses to entice a few. For most students, the project is about learning how to: (1) work with others, (2) manage a project, (3) think critically and scientifically, and (4) communicate both verbally and orally.

Because this project will be used either directly or indirectly by the student when seeking employment later, the student should select a research project that is either personally compelling, practically valuable, or both. A personally compelling question will bring the student to the work for pleasure rather than working on just another assignment. A practically valuable project in which publication is likely or the skills are directly conveyed for future work offer external rewards that appeal to some. The combination is the ideal, as this offers both intrinsic motivation to seek the answer and conduct the work and extrinsic rewards that are easy to use as goalposts. Regardless, bad project choices can lead to very disappointing outcomes at the end of the term. Motivation makes hard work easier or at least less onerous.

COLLABORATIVE REPLICATION AND DIVERSITY, SOCIAL JUSTICE, AND SUSTAINABILITY

In Grahe et al. (2019), we make extensive arguments about how adopting open science initiatives benefitted diversity, social justice, and sustainability objectives. In this section, I will briefly recall the arguments about why collaborative projects, like the CREP and NICE, specifically benefit diversity, social justice, and sustainability objectives for science as well as the students and faculty who conduct the research. These two projects, in particular, offer very good examples of the values of these undergraduates, which is a different collection of researchers compared to graduate students and doctoral-level researchers. However, the benefits generalize well to other projects such as the PSA or Many Labs projects, and they offer different diversity, social justice, and sustainability advantages that should be considered in more depth elsewhere.

Projects with many researchers are, by definition, more diverse than projects with fewer researchers. When projects invite researchers from anywhere in the world to share research protocols, the researchers and populations they measure will be more diverse than a solo researcher repeatedly sampling from the same population of students. How much of psychology differs across populations separated by geography, language, religion, and culture? We really do not know. The vast majority of research has been conducted on Western, educated, industrialized, rich, and democratic (WEIRD) populations (Henrich, Heine, & Norenzayan, 2010). While there are examples of studies sampling broadly across many countries, any one of these studies should also be replicated repeatedly to draw strong inference. Further, by limiting the research questions to those from WEIRD populations, we might not be asking the right questions about psychology from populations that are not WEIRD. Crowd projects that invite anyone to join encourage more diversity. Both the CREP and the NICE actively sought to coordinate with researchers outside the United States even though they were founded in the US. The PSA requires a portion of their leadership to be international to guarantee the representation of more diverse voices.

Does increased diversity benefit anyone? Some might argue that seeking diversity is distracting from the research process. However, evidence suggests that students from disadvantaged populations are less likely to seek out research opportunities on their own (Bangera & Brownell, 2014). One benefit of using the CREP or NICE is that all students in the class are introduced to the value of participating in research. Further, all the students can earn the rewards associated with being a part of these projects, such as conference presentations, co-authorship on manuscripts, and networking with contributors at other locations. In this sense, these crowd projects offer greater social justice. Not only are the student researchers receiving more social justice, but their instructors have access to greater research resources.

There are schools that demand more research output, and there are schools that have more resources. When schools have both, researchers have plenty of resources to conduct their own work (e.g., Harvard or University of Michigan). Crowdsourcing might be less appealing to many in this situation. In contrast, crowd projects are ideal for situations in which there are few resources and modest research expectations (such as my own institution, Pacific Lutheran University) or those with no resources and no research expectations (such as most two-year community colleges). For faculty and student researchers at these schools, crowd projects are inviting because of the shared resources and because the rewards are commensurate with the institution's expectations. Having identified the other categories, I must imagine that some schools offer lots of resources but low expectations for output, though I am not familiar with any examples. In every case, the researchers are welcomed into the crowd and invited to share the effort to answer important questions.

Faculty and student researchers in situations like mine and those worse off experience many resource challenges. One of the biggest is obtaining reasonable data to satisfactorily draw an inference. Human participants are limited, and the populations are geographically homogenous if not culturally so. By collecting data from multiple locations simultaneously, the research addresses both these problems. The sample size is finally large enough to test hypotheses sufficiently, and the samples generalize beyond the small populations they were sampled from.

One of my favorite memories of this was when a faculty member at a tiny, geographically isolated school was asking me for advice on how to conduct an analysis on EAMMi2 data. The data set was so big and complicated that her students had to do an analysis she had never before needed. "My data sets are just never this big or complicated," she exclaimed. My response was, "You're welcome. I built this data set for people just like you." I have returned to that line many times with faculty at small institutions across the country who have worked on papers with me or on their own. These data are just more valuable than what they can access normally, and they are very happy to use them. Better data means better publishing opportunities. Crowd projects are more publishable with their better data and are cited more frequently for the same reason. Having more publishing opportunities for individuals from disadvantaged economic backgrounds is a major contribution toward social justice.

Another main contribution is that the findings apply more broadly and benefit more people who were not previously represented and are likely to be economically disadvantaged. When considering diversity, I mentioned this in the context of collecting better data by reaching diverse, non-WEIRD populations that were previously not measured. Also, these researchers were not included in decisions, and their questions were not included in the pursuit of science more broadly. From a social justice perspective, the data drawn from studies that include local researchers and broader populations can now be applied to and benefit those previously underserved populations.

Often, when people think about sustainability, they think about nature and environmentalism. I agree that sustainability goals should always consider the resources supporting the outcomes. If I were in charge of more resources, I would make the United States a very nature-friendly landscape. I hope that a few readers will increase their own passion to improve the natural environment. However, in this context, sustainability refers directly toward the research itself. When considering more diverse researchers and populations generating better studies that are more generalizable across more people, sustainability is seen in the duration of our understanding of truth.

A gentle reminder: science uncovers the best version of truth we have at this time. Even for things like gravity, a scientist should acknowledge that we still do not know the complete truth. For something such as psychology, in which individual differences are larger than group differences in almost every setting, our current understanding of truth should never be considered permanent. Moreover, we are evolving culturally all the time. Thus, truth for psychology is ever changing in some facets such as communication and information processing, to name just two from a possibly infinite list. Thus, to broaden and extend our understanding of scientific truth is critical, and I argue that these projects do just this for our science and make them more sustainable.

Psychological research, particularly applied research, is intended to improve the lives of humans. However, horrific research has been conducted to injure the lives of others in the name of science. While researchers like to think only about evil Nazi or other nefarious enemy scientists, two US clinical psychologists were integral in helping develop the torture procedures that were decried by humanitarians throughout the world, and they were admonished by the APA (Pope, 2015). However, the purpose of the research is to better people. Here, I remind you that the researcher is human too. Broader psychology uplifts humanity by helping us achieve our best selves and by minimizing pain caused by mental illness, death, trauma, and other horrors of existence.

But systems need to be in place for the people who conduct that research. Not all instructors have time to conduct research, have resources from their institutions, or have much intrinsic interest. Yet most instructors have some responsibility to help students learn how to conduct research. Further, there are faculty who like research but cannot conduct any because they are teaching students. In a given year, research methods or capstone instructors might mentor 5, 10, 20, or even more student research projects in a year. Because of the inherent challenges of single studies that are consistently underpowered in topic areas that are diverse and tangential to the instructors' personal area of interest, it is exceedingly rare for these studies to amount to publications.

Personally, this cycle and impact on my own work caused me to almost leave the profession. What saved me was this idea to collect data across campuses and two role models who said, "Jon, that is a great idea!" Since then, many projects have allowed instructors and their students to publish research. More importantly, even the data that are not published are documented in public spaces so that future researchers might find them. In fact, one of the first studies that the CREP included for contributors (Elliot et al., 2010) was found by a meta-analytic researcher who was scouring the internet looking for replications of that study. The students' publicly accessible data were found and included in that meta-analysis (Lehmann, Elliot, & Calin-Jageman, 2018) as well as a paper reporting the complete set of CREP samples (Wagge et al., 2019).

Some readers right now might be thinking about research questions that could be answered by accessing all of the CREP samples (> 100 from 13 studies) and conducting some metascience meta-analysis (e.g., did effect size differ by location of contributor or type of school? Do CREP pages improve over time?). Regardless of the question, whether it relates to the theoretical question from the original research or from the way that the contributors replicated work that students and their instructors conducted and then discarded without further consideration is now being documented and offered in permanent repositories. That represents sustainable science contributions.

Chapter 2: Crisis Schmeisis Journey Book Review

African American Women of the Old West

Tricia Martineau Wagner

This book is about 10 women who journeyed West during the 19th century. Their stories are all very different and inspiring. It fits this chapter because it shows the diversity of these women even though at the time they were all labeled similarly by the culture around them (https://osf.io/tz6e7/).

CHAPTER 2 EXERCISES

Exercise 2.1: Estimating Conditions That Impact Data Collection

- Make a list of predictor variables, subject variables that are physical (body strength), motivational (hiking as a passion), and context driven (privileged childhood).
- Make a list of measurement variables related to the location (place in park), researcher (appearance), or other conditions (timing).
- Estimate how differences could change the outcomes.
- Consider your own research and what variables might impact outcomes and should be controlled.

Exercise 2.2: Comparing Replications

- Choose one of the published papers from CREP data.
- Read the original paper and the CREP replication.
- Answer the 36 Replication Recipe Questions for the CREP paper.
- Brandta et al., 2014
    - www.sciencedirect.com/science/article/pii/S0022103113001819
- CREP original paper—CREP replication paper

- Forest & Wood, 2012; Leighton, Legate, LePine, Anderson, & Grahe, 2018
- Elliot et al., 2010; Wagge et al., 2019
- Eskine, Kacinik, & Prinz, 2011; Ghelfi et al., 2020

Ghelfi, E., Christopherson, C. D., Urry, H. L., Lenne, R. L., Legate, N., Ann Fischer, M., . . . Sullivan, D. (2020). Reexamining the effect of gustatory disgust on moral judgment: A multilab direct replication of Eskine, Kacinik, and Prinz (2011). *Advances in Methods and Practices in Psychological Science*, *3*(1), 3–23.

Exercise 2.3: Expanding the Literature Review

- Finding the right articles might be the most difficult first step in conducting research after an idea is identified.
- The American Psychological Association offers this *general advice*, which is quite useful: www.apa.org/education/undergrad/library-research.
- Of course, most of us are a bit more sophisticated now, and for that, *APA offers this excellent introduction* to conducting a good literature review: www.apa.org/science/about/psa/2013/10/using-psycinfo. For students who are struggling to find relevant articles, this is an important resource to review.
- Beyond PsycInfo, two other search engines that could also be employed whenever you are writing a paper for publication are http://Scholar.google.com and PsyArXiv Preprints.
- Store copies of manuscripts you find on your background OSF component, and make sure to keep that component private. Because of copyright issues, it is OK to store published work on the OSF, but it cannot be shared publicly.

END-OF-CHAPTER REVIEW QUESTIONS

1. Which three parts of speech can "replication" act as?
2. Explain the three types of replication and how they are different.
3. What makes a replication successful?
4. Name three factors to consider when planning research in a national park.
5. How does collaborative replication support diversity, social justice, and sustainability?
6. What are some potential advantages to performing replication science? Can you think of any disadvantages and possible solutions to help mitigate any disadvantages?
7. Assuming appropriate funding, think of a research question that can be easily replicated within the National Park Service. What aspects of the research question lend themselves to replication?
8. Name three different crowdsourcing or replication projects described within this chapter. Compare and contrast them.

References

Anderson, S. F., & Maxwell, S. E. (2016). There's more than one way to conduct a replication study: Beyond statistical significance. *Psychological Methods, 21*(1), 1.

Bangera, G., & Brownell, S. E. (2014). Course-based undergraduate research experiences can make scientific research more inclusive. *CBE—Life Sciences Education, 13*(4), 602–606.

Brandta, M. J., Izermana, H., Dijksterhuis, A., Farach, F. J., Gellerd, J., Giner-Sorollae, R., . . . Van't Veera, A. (2014). The replication recipe: What makes for a convincing replication? *Journal of Experimental Social Psychology, 50*, 217–224.

Cumming, G. (2014). The new statistics: Why and how. *Psychological Science, 25*(1), 7–29.

Cumming, G., & Calin-Jagerman, R. (2016). *Introduction to the new statistics*. London: Routledge.

Elliot, A. J., Kayser, D. N., Greitemeyer, T., Lichtenfeld, S., Gramzow, R. H., Maier, M. A., & Liu, H. (2010). Red, rank, and romance in women viewing men. *Journal of Experimental Psychology: General, 139*(3), 399–417.

Eskine, K. J., Kacinik, N. A., & Prinz, J. J. (2011). A bad taste in the mouth: Gustatory disgust influences moral judgment. *Psychological Science, 22*(3), 295–299.

Forest, A., & Wood, J. V. (2012). When social networking is not working individuals with low self-esteem recognize but do not reap the benefits of self-disclosure on Facebook. *Psychological Science, 23*, 295–302.

Fosse, N. E., & Toyokawa, T. (2016). Interinstitutional variation in emerging adulthood: Does selectivity matter? *Emerging Adulthood, 4*(3), 142–152. doi:10.1177/2167696815585685

Ghelfi, E., Christopherson, C. D., Urry, H. L., Lenne, R. L., Legate, N., Ann Fischer, M., . . . Sullivan, D. (2020). Reexamining the effect of gustatory disgust on moral judgment: A multilab direct replication of Eskine, Kacinik, and Prinz (2011). *Advances in Methods and Practices in Psychological Science, 3*(1), 3–23.

Grahe, J. E., Corker, K. S., Schmolesky, M., Alvarez, L. D. C., McFall, J., Lazzara, J., & Kemp, A. H. (2020). Do institutional characteristics predict markers of adulthood?: A close replication of Fosse and Toyokawa (2016). *Emerging Adulthood, 8*(4), 270–284. doi:10.1177/2167696818810268

Grahe, J. E., Cuccolo, K., Leighton, D. C., & Cramblet Alvarez, L. D. (2019). Open science promotes diverse, just, and sustainable research and educational outcomes. *Psychology Learning & Teaching, 19*(1), 5–20.

Henrich, J., Heine, S. J., & Norenzayan, A. (2010). The weirdest people in the world? *Behavioral and Brain Sciences, 33*(2–3), 61–83.

Ioannidis, J. P. (2005). Why most published research findings are false. *PLoS Medicine, 2*(8), e124.

Lehmann, G. K., Elliot, A. J., & Calin-Jageman, R. J. (2018). Meta-analysis of the effect of red on perceived attractiveness. *Evolutionary Psychology, 16*(4). doi:10.1177/1474704918802412

Leighton, D. C., Legate, N., LePine, S., Anderson, S. F., & Grahe, J. (2018). Self-esteem, self-disclosure, self-expression, and connection on Facebook: A collaborative replication meta-analysis. *ResearchGate*. doi: 10.24839/2325-7342.JN23.2.98

Maxwell, S. E., Lau, M. Y., & Howard, G. S. (2015). Is psychology suffering from a replication crisis? What does "failure to replicate" really mean? *American Psychologist, 70*(6), 487.

Moshontz, H., Campbell, L., Ebersole, C. R., IJzerman, H., Urry, H. L., Forscher, P. S., . . . Chartier, C. R. (2018). The psychological science accelerator: Advancing psychology through a distributed collaborative network. *Advances in Methods and Practices in Psychological Science, 1*(4), 501–515.

Patil, P., Peng, R. D., & Leek, J. T. (2016). What should researchers expect when they replicate studies? A statistical view of replicability in psychological science. *Perspectives on Psychological Science, 11*(4), 539–544.

Pope, K. S. (2015). Are the American psychological association's detainee interrogation policies ethical and effective? *Zeitschrift für Psychologie, 219*(3). doi:10.1027/2151-2604/a000062

Spellman, B. A. (2015). A short (personal) future history of revolution 2.0. *Perspectives on Psychological Science, 10*(6), 886–899. doi:10.1177/1745691615609918

Uhlmann, E. L., Ebersole, C. R., Chartier, C. R., Errington, T. M., Kidwell, M. C., Lai, C. K., . . . Nosek, B. A. (2019). Scientific utopia III: Crowdsourcing science. *Perspectives on Psychological Science, 14*(5), 711–733.

Wagge, J. R., Baciu, C., Banas, K., Nadler, J. T., Schwarz, S., Weisberg, Y., . . . Grahe, J. (2019). A demonstration of the collaborative replication and education project: Replication attempts of the red-romance effect. *Collabra: Psychology, 5*(1).

3 PREREGISTERED

Determining Answers to Decisions Before They Happen

Chapter 3 Objectives

- Define preregistration and understand its purpose
- Consider different types of preregistration
- Understand the costs and benefits of preregistration
- Consider relationships between diversity, social justice, and sustainability and preregistration
- Understand the content to be included in preregistration
- Understand when to preregister
- Review preregistration and undergraduate research projects

MUSIC IN A BOX: ABOUT THE SONG "PREREGISTERED"

Another open science initiative to increase transparency is the process of preregistering and documenting hypotheses, methods, and analysis plans. This song explains all the decisions researchers need to make, freeze, and declare before data collection. Conveniently, the song also follows the Center for Open Science Preregistration template, helping researchers answer the questions posed on the form. This song has a fast moving pace and changes chords quickly with up-tempo urgency.

DEFINING PREREGISTRATION

Preregistration, very broadly, is the process of documenting the research plan before the data are collected. This is one of the most radical changes to emerge from the Scientific Revolution 2.0, and the processes and expectations for what makes a good preregistration are slowly emerging. What makes this so radical is the change in when the researcher writes and reports on the methods. In the time just before the crisis, many authors endorsed Bem's (2000) arguments that the researchers' should not bother the reader with errors in conducting the project. Instead, an introduction and methods should be crafted to present a narrative that reflects the data collected. The problem with this approach to writing occurs when the author presents exploratory findings as confirmed hypotheses. Besides the glaring misrepresentation of the purpose of science, this practice encourages people to take advantage of a priori tests, even though they should be using post hoc tests. The reader might recognize the description of HARKing.

This process reflects questionable research practices in making decisions regarding **researcher degrees of freedom** (Wicherts et al., 2016). Researcher degrees of freedom reflect any decision that a researcher might make during the process that could have been decided differently. Wicherts et al. describe 32 different places across a manuscript where these are present. Preregistration is an effort to freeze, or predetermine, those decisions before the data are collected by requiring the researcher to document the decisions and then date-stamp them using a permanent repository. Chapter 5 will describe some examples of questionably using researcher degrees of freedom. Noting their existence here serves to explain why we need preregistration. Researchers, flawed by humanity, forget or experience biased recall. As the writing process can take years, returning to some detail that needs further elaboration later in the process can be laborious. Preregistration tries to limit these unintentional errors as well as the more intentional misstatement of the research process.

Preregistration Can Be Used for Both Confirmatory and Exploratory Hypotheses

Preregistration helps a researcher demonstrate which hypotheses were truly confirmatory by date-stamping them. More than one prominent researcher has said something such as, "All my research is exploratory anyway. There is no point for me to preregister." However, this approach assumes that the preregistration is for some third party who only wants to track their honesty. Instead, preregistration offers the researcher an opportunity to "front-load" the writing process and reduce challenges of miscommunication between collaborators during the collection and analysis steps. For confirmatory hypotheses, researchers should document as much detail about the process as possible. This will avoid later criticisms or the temptation to find reasons to remove data that do not fit the hypotheses. For exploratory questions, researchers want to document the process in as much detail as possible to increase the ability to detect data of interest by identifying methods for the analysts to use when the data are examined. In both cases, the researcher should think deeply about the materials and procedures that will generate the data and how the data should be collected, managed, and analyzed.

In either case, this process of registering a page freezes the contents at that time such that the owner can no longer edit it in the future. It does not restrict a researcher from future action, but it does provide evidence of what the researcher was intending at that time. These can be as simple as very broad statements written onto an OSF wiki with no intended methods. For instance, one could preregister a hypothesis such as, "I predict that either a Republican or a Democrat will win the next United States presidential election." It is a very safe prediction, since the last time this was not true was in the early 1800s. One could be bolder and predict who might be the winner of that election or of any other. I could predict my research hypothesis posed in the previous chapter of this book, "I predict a positive curvilinear relationship between the age of visitors to national parks and the miles they walked." In fact, the book will use this example for the remainder of this chapter to highlight how and what should be preregistered in the design and methods as well as the intended analyses.

TYPES OF PREREGISTRATION

There are three ways to use preregistration, and there are many preregistration formats that can influence how a paper is written. This section will briefly contrast the three ways they are used and then introduce two of the formats while directing the reader to further resources. The differences reflect motivations of the researchers and methods of conducting research or approaching analyses. There is no best format, though there are many poorly constructed preregistrations from early attempts of unequipped researchers who were not yet ready to fully specify their questions and process. For those students who are reading this book as a requirement and not looking toward a research career, preregistration might feel like busy work for their project. However, the decisions need to be made, and when researchers answer the questions earlier rather than later, they save time writing and avoid potential errors in data collection. For students who have interest in presenting their project at a research conference, sharing it with a larger group (i.e., CREP or NICE), or writing it up for publication, the preregistration format is critical. And for those intending careers in research, mastery of the preregistration process will enable deeper understanding of all facets of the research process.

The biggest difference in how preregistration is used is whether the preregistration is submitted to a journal before the data are collected. As a researcher, I have regularly gotten my research prepared to conduct, received IRB approval, preregistered my work at that time, and then started data collection. Later, when I submitted my work for publication, I noted that it was preregistered as part of the justification for the confidence in the findings. This type of preregistration does not allow for flexibility of design by the editors and reviewers to potentially address research flaws before the data are collected.

In contrast, a **registered report** is a type of manuscript where the authors submit a **stage 1 preregistered report** to a journal and receive feedback on the Introduction, Methods, and intended Analyses before collecting any data. The authors and editors interact one or more times through reviews and resubmissions until the authors and editors agree in principle to publish the outcome of the study regardless of the statistical significance of the results. An **in-principle acceptance** requires the researchers to demonstrate that they followed the protocols they agreed to. The researchers then collect their data, follow analysis protocols, write the results and discussion, and submit a **stage 2 preregistered report** for final review and likely acceptance. For **Registered Replication Reports** (RRRs), in addition to being replications rather than novel research like Registered Reports, the goal is to collect as many other researchers from as many different laboratories as possible.

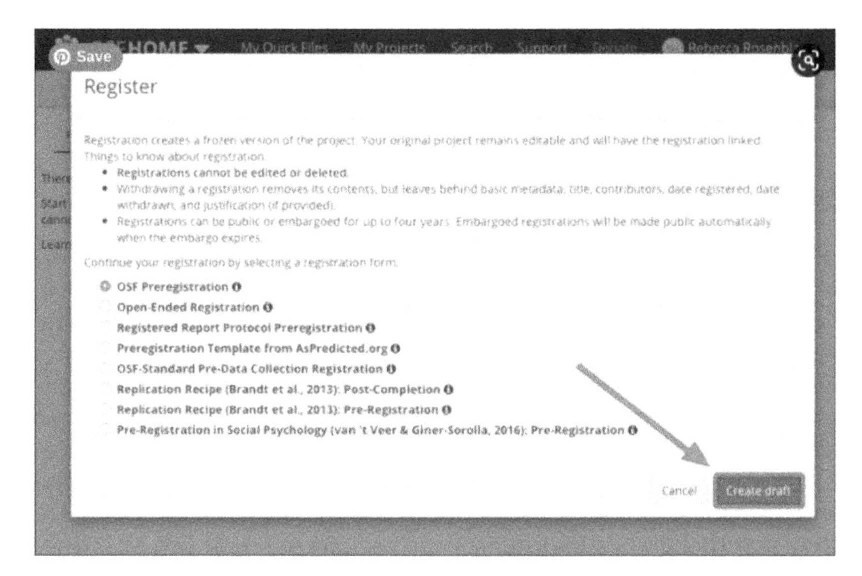

Figure 3.1 Preregistration choices in the OSF

The journal *Psychological Science* has published a number of these massive endeavors. They are large-scale, crowdsourced projects that occur for a single question rather than an ongoing organization. For instance, the CREP and the PSA teamed up to create the Accelerated CREP, which was a merger between the two organizations to conduct an RRR testing the reproducibility of Turri, Buckwalter, and Blouw (2015). Once the paper (Hall et al., 2020) obtained an in-principle acceptance in the stage 1 step, researchers began collecting data following an agreed-upon protocol. During the pandemic of 2020, many labs interrupted data collection, and the project was stalled. However, dozens of teams had already completed their work. To quickly summarize, anyone can preregister any project. If they want editorial feedback on a journal submission before they conduct the data collection, authors write a registered report. If the researchers are focused on a single effect and intend to conduct a massive large-scale test of it, they might seek to conduct an RRR.

HOW PREREGISTRATION BENEFITS THE RESEARCHER IN THE SERVICE OF SCIENCE

In scientific research, like life, there are costs and benefits to everything. Adopting open science practices does have some labor costs, particularly considering the learning curve. And yes, the benefits in the outcomes are intended to exceed the costs. I do not offer my students busywork, only work to help them learn, and I extend that notion here. This commitment to open science practices is intended to benefit the researcher individually in the service of science and beyond. This section briefly focuses on the benefits to the researcher of preregistration and introduces four principal benefits: (a) "front-loading" workflow to before data collection, (b) reducing self-deception, (c) making decisions before they matter, and (d) obtaining clear evidence of a priori capacity.

Front-Loading Work

The practice of preregistration is identifying the hypotheses, plans, procedures, materials, data management, and data analyses before data are collected. All of this work should happen regardless of the outcomes of the data analyses. In the past, the documenting of this work normally happened after the data analyses. Much of the timing was related to the way researchers wrote papers before computers, because if the data turn out to demonstrate that there is no effect, most researchers just deposited their work in a file drawer. I won't bore the reader with stories of storing data/materials/analyses/papers in ancient metal file cabinets. But manuscripts deemed to have failed because the research reported did not find statistically significant results lined the drawers of most researchers' file cabinets, my own included. Science has errors.

The thing is, we are supposed to document those errors in order to provide evidence of where science does or does not succeed. Of course, the problem is that no one would publish a "failed replication" because it was often attributed to be a problem of the researcher doing the replication rather than the original finding. Once an effect or theory gets into textbooks, it is difficult to convince many that it is not true. However, some effects seem wrong or at least not fully explained (e.g., ego depletion, facial feedback hypothesis, embodied cognition of power poses). By documenting what researchers do along the way and making it easy to share findings regardless of outcomes, science benefits through more accurate and complete information. For the researcher, the work is moved from after analyses to before. This benefit is tied to the second benefit, reducing self-deception.

Reducing Self-Deception

People cheat and are dishonest. I have seen minor cases throughout my career that do not warrant the news focus of the one we discussed in Chapter 1. I have been told by students about cheating in my courses from time to time. Sometimes people cheat, and no one can stop that, but the rest of us should trust each other. However, in science, we can cheat without knowing we are doing it. It all comes down to the reliability of the probability that a statistic is unlikely to have occurred by chance, the p-value. This is discussed in greater detail in Chapter 4 and was hinted at in Chapter 1, but the crux of the problem is that each time we look at the same data set, the true p-value is inflated relative to the one offered by computer software or the tables in the back of a book. So . . . we cannot trust ourselves.

I will describe a very honest and erroneous situation that changes the p-value. Pretend that we have collected data on age and number of miles walked by visitors from each national park. We run our analysis and find no significant effect. Mind you, we just spent thousands of hours planning and executing this study and thousands, if not hundreds of thousands, of dollars to collect the data. If the planned analyses turned out to demonstrate that there were no effects, should the data be tossed in a file drawer to be forever ignored? Maybe there is some bad data from bad data entries or bad subjects, and if they were deleted, the effect would emerge. Maybe, if the national parks were categorized by region or population density or average temperature or cost of surrounding communities, we would see the effect. How many analyses should be conducted to find a true effect within the data that cost us years and so many resources to collect? Personally, I would conduct many analyses looking for something to salvage.

The problem is that each time one of those new approaches is tested, the true p-value, and therefore the associated alpha measuring Type I error rate, increases. The inflation rates differ depending on the practice employed, but the process is the same. Each time we look at the analyses again, the error rate increases. There are also many corrections to that value that might be employed (such as Bonferroni or Sheffé tests). However, remember that by 2010, a dominant practice was to rewrite the paper as though the one significant effect was the planned analysis. In this case, no correction to the p-value is needed. The reader might think that authors were being willfully deceptive, but rather, they should assume that the authors were forgetting the many decisions they made. Preregistration helps remind the researcher of each of these decisions before the data occur. Then the researcher can be clearer with the difference between an exploratory question, which needs correction to the p-value, and confirmatory research, which does not.

Making Decisions Before They Matter

Related to the previous two concerns, if preregistration requires decisions before data collection occurs, and if the process helps us remember those decisions, this early process offers some protection against methodological errors. These are decisions about which measures to use, how many subjects are needed to achieve adequate power, how to handle missing or bad data, or other data management issues. In the process of making decisions, the researcher can imagine the experiences of the subject, practice with simulated data sets, even conduct pilot work and examine actual data that will not be in the final report. If errors are identified here (e.g., missing items, underestimated effect sizes), then changes can be made before the costs of data collection are expended.

Clear Evidence of A Priori Capacity

Researchers' careers are advanced more when other researchers read and use their work. Those of us motivated by the need for achievement are looking for that idea or finding that is so important that news of it reverberates across the field and beyond. More will be discussed about measuring success in Chapter 7, but recognizing that researchers want to be discoverable helps us see a benefit of registration. By documenting the hypothesis, materials, and procedures of a study, researchers make it easier for others to use their work. When the data are added later, reanalyses of secondary data by meta-analysts, methodologists, or others in the same subdiscipline also offer opportunities for research to have more impact.

However, the real value of preregistration is evidence of a priori (beforehand) knowledge. Hypotheses that researchers predict are better than ones they discover. It suggests their theoretical insight is special. Recall that one cause of the Replication Crisis is that the publication process had begun to encourage researchers to present their post hoc (after-the-fact) findings as though they were a priori. Finally, there was little oversight or concern about how researchers removed data that were considered bad. All of this inflated the Type I error rate in published research with reported effect sizes that are larger than reality because of artificial data cleaning procedures. With a preregistered design and analysis plan, researchers can confidently present a priori findings that reviewers and editors will not question based on questionable practices with researcher degrees of freedom. In contrast, the researcher degrees of freedom are now "frozen in time" on a repository that cannot be altered with a date stamp that precedes data collection.

DIVERSITY, SOCIAL JUSTICE, AND SUSTAINABILITY AND PREREGISTRATION

Stephen Lindsay, an open science champion, was criticized for a paper that was published in a journal when he was the editor (Clark, Winegard, Beardslee, Baumeister, & Shariff, 2020). The manuscript made claims about religious beliefs, intelligence, and likelihood to commit violence and found differences between poor and rich countries with different racial makeups. Lindsay wrote a strong apology for not recognizing that some of the methods and analyses were less rigorous, and later, the authors conceded that they were not as careful as they should have been. My own response was relief that I did not face the same level of criticism and scrutiny for manuscripts published in the journal of which I am editor. Of course, with more transparency in data and materials, it is easier to check the veracity of researchers' claims and if this work had been preregistered, they might have realized that their metrics were biased toward WEIRD populations. In short, predictions about race, culture, gender, and other inflammatory topics should be demonstrably "a priori." Finding differences post hoc invites explanations that reflect personal biases rather than reality.

To emphasize this point, I tell a brief story. While in graduate school, I was privileged to be added as fourth author on a paper with my new advisor about what behaviors predicted dyadic rapport across situations (Bernieri, Gillis, Davis, & Grahe, 1996). In the data, there was a statistically significant, moderately sized correlation between dyad race similarity and dyadic rapport.

In other words, if both members of the dyad were of the same race, they felt more rapport with each other than if they were of different races. At the time, we had no explanation for this. It certainly was not predicted, but the research was exploratory and required us to measure as many aspects of the dyad as we could. My advisor was very clear that we should not try to explain this effect, as it could be a statistical anomaly (a Type I error), because our sample size was not that big, and we had no a priori explanation.

Years later, I think it might be that this effect might be real and I think it can be explained by the presence and perception of microaggressions (Sue, Capodilupo, Nadal, & Torino, 2008) during the interaction. Though I do not have adequate resources to test this prediction, I could preregister it and even write a paper recommending ideal methods of testing this prediction. There are others who might have these resources if I can justify my argument using their proposal system, such as the PSA or the NICE. Predictions about inflammatory topics should be demonstrated a priori and tested with adequate power to make convincing inferences. In this way, preregistration protects diversity by not allowing bad arguments to stand.

If you ever meet me in person and ask me the right questions, I'll tell you about some ideas I feel others might have scooped after hearing them from me. Of course, sometimes, our ideas are just good enough that other people have them too. The problem with ideas is that they have no value unless they can be connected to outcomes. An example of an idea that multiple people had occurred back in the early 2000s as the internet became more usable. I told a colleague that I had better ideas than resources and wanted to publish research questions that I would like to see conducted in hopes of finding collaborators. His response was that my idea was crazy and that no one would do that. Of course, Study Swap (https://osf.io/meetings/studyswap/) is exactly this concept. I never acted on that idea because my colleagues' negative outlook interfered with my progress. However, the reality of that idea offers many of us with fewer resources the capacity to answer questions we were trained to answer.

Both the PSA and NICE accept proposals to conduct crowdsourcing research. Further, RRRs offer one-off opportunities to find large groups of researchers to test the reproducibility of a single (or a couple) effect. Finally, Study Swap offers small collaborative opportunities for this where people with needs (an idea that needs tested) find people with capacity (extra subjects) or an interest in the collaboration. In short, researchers who are underresourced, whatever the reason, can now connect their ideas to outcomes. By documenting and date-stamping an idea, they can share it with potential collaborators, who cannot steal it because there is a permanent record of who had it. If they do answer the question on their own without the person who generated the idea, now they need to acknowledge the prior date and the source of the idea. That is social justice.

As diversity and social justice are enabled by documenting ideas before they occur, sustainability is advanced, because findings derived from confirmatory research have more lasting impact than those from exploratory findings. Because the nature of exploratory research is about discovery, there is more tolerance for Type I errors, as Type II errors are being avoided. The goal is to find an effect and then invite the science world to test it through replication. All truth is temporary and relative, and research is trying to make a statement about truth that will extend the longest. Confirmatory research tests findings with high power to reduce both Type I and Type II error while also reducing Type I errors with more severe corrections to p-values. The process of preregistration documents more effectively which effects are confirmatory and which are exploratory, thus helping the readers estimate their own perceptions of the sustainability of an effect.

NATIONAL PARKS IN FOCUS: CONNECTING PREREGISTRATION QUESTIONS TO VACATION PLANS IN NATIONAL PARKS

Have you ever had a vacation with the goal of visiting multiple stops, maybe multiple national parks, in the same trip? The National Lampoon's *Vacation* movies with Chevy Chase offer a comedic approach to such a trip. In each movie, the vacation plan is repeatedly interrupted or derailed with some distraction or trauma. That discrepancy between plans and reality very often reflects the experience of research. Anyone who has prepared a study and collected data can identify with that moment when the first results emerge in the output file after hitting the [submit] button. There is either this moment of elation when the numbers match hope or despair if no effect emerges at all. To try to connect the idea of preregistration more closely to research, this section extends our national parks metaphor.

Imagine two vacationing families with opposite approaches to a trip to a series of national parks over two weeks. These two families plan the same "car tent camping" trip for the same time of year and live in approximately the same location. They intend to go and put up a tent in the designated park camping grounds. Imagine they are leaving from their home to drive 15 to 20 hours to arrive at the first park. After two to three days, they will go to another and so on until they drive 15 to 20 hours home. With 4 days of travel to and from the first destination, they can visit three or four parks depending on distance (e.g., Yellowstone, Grand Tetons, Rocky Mountain National Park, and Grand Canyon).

One family, the Confirmers, is extremely organized and starts planning a year in advance. They have all their lodging reserved and a schedule of activities at each location to make sure they enjoy each moment in each park to the fullest. The other family, the Explorers, does not plan beyond identifying the dates of their trip and the materials (transportation, clothing, food, tents). They intend to find camping and leisure activities when they arrive in the area. I wonder, as this scenario is being described, how many readers identify their families or themselves as Confirmers or Explorers. For those of us who have tried a vacation like this, we might also be imagining possible outcomes for these two families. If one or the other sounds more like trauma than vacation, I will offer that each family might find their vacation ruined depending on the context of their trip.

While considering these two imaginary families, try to imagine all the other families on this planning continuum. In this exercise, consider how conditions for each of these families might impact both their enjoyment of the trip and our DV of interest (miles hiked).

Reading Exercise—How Conditions Impact Enjoyment and Miles Hiked

Conditions

- No open spaces for camping in the park campground
- Park is busy, and planned activities are delayed by traffic jams or long lines
- Rain, snow, fire, and other natural disasters
- Car breaks down
- Children disagree with plans

Responses

- What is the likely response for each?
- How would they impact the family fun?
- How would they impact the number of miles hiked by each member of the family?

Ideally, this exercise highlighted that in some cases, the Confirmers would likely have a better time, while in others, the Explorers would have the advantage. Each type of research is important. Explorers have more flexibility to address some unexpected challenges and obstacles, since they could change all their trip plans to avoid unexpected tragedies like major fires. On the other hand, Confirmers have more security in their plan, since finding solutions to overfilled campgrounds can be time consuming and expensive. What other advantages or disadvantages did you find when considering these extremes in planning?

WHAT AND WHEN TO PREREGISTER

This section will introduce the basics of the preregistration process and advise when to preregister a research project. I also offer a completed sample preregistration using our book hypothesis (positive curvilinear relationship between age and miles hiked). Readers who are getting ready to preregister their own work should review the questions and think deeply about them before completing a preregistration. Remember, this is not busywork; it is critical work. To sufficiently answer these questions, a researcher needs to clearly and intricately understand their study and materials. Flake and Fried (2020) offer six questions that researchers should answer before they execute their research in order to refine the transparency of the study. For each question, they note types of information that should be included in the answer. Before considering the questions that will be asked in a preregistration, consider these questions about your research.

Flake and Fried (2020) ask authors to start by first answering, "What is your construct?" Here, the author transparently defines the construct and describes theories and research supporting the construct. Next, they direct authors to answer the question of, "How do you operationalize your construct?" Authors should answer this by describing the measures and procedures while matching the measure(s) to construct(s). Once the construct is sufficiently defined, they suggest authors should explain, "Why do you select your measure?" Authors need to justify their measure selection and report any evidence of validity. These first three questions are more conceptual but still require precision for rigorous science.

The next two questions regard specifying the measures' computation: "How do you quantify your measure(s)?" and "Did you modify the scale?" To address the quantification question, authors need to first describe all the item responses and transformations, items per score, item scoring, and all parametric analyses computed. To answer modification questions, authors need to describe any changes made to the scale, when they occurred, and why the decision was made. Finally, Flake and Fried ask authors to answer, "Did you create the scale on the fly?" If the scale was built for this study, justify why an existing scale was not used, report all the parametric details, and provide any evidence for validity of the construct.

These questions invite authors to more fully consider all measurement aspects of a study. If researchers plan a study fully, all these questions will be answered in the planning. Ideally, an Introduction and Methods section would offer all this information, but this is not always achieved. Even when preregistering, authors can present too little or too vague information. Take a few moments and review the constructs, measures, and procedures of your study. Then proceed with learning about the content needed to preregister your research.

Confirmatory Research Questions

The reader has probably guessed by now that every possible decision known before the study is conducted needs to be preregistered, including the direction and expected magnitude of the hypothesized effect size, how variables will be operationalized, how data will be collected, how many subjects will be needed, how data will be managed, and how analyses will be conducted. These broad questions contain many smaller questions. Recall that Wicherts et al. (2016) identified 32 distinct researcher degrees of freedom in any research report. For confirmatory research, each of these researcher degrees of freedom needs to be eliminated.

Exploratory Research Questions

Because preregistration for exploratory research does not get any statistical power boost from preregistration, explorers sometimes claim the process has no value to their work. However, as I noted earlier, the value is really to document work for future selves or collaborators so that there is less confusion and fewer errors. Decisions that are known before data collection should be documented. A justification for the study and the methods of data collection should both be known before the study even if there are no expected outcomes. Also, the first attempt at data coding and data analysis are likely identified by this point. Researchers might also want to note next steps if findings come out a certain way. In the COS Preregistration form, researchers are asked about possible follow-up analyses as an optional question. While not necessary, this can be shared if known.

When to Preregister

For any study that needs to be preregistered, there are two occasions where this process should occur: before data collection and after data cleaning and analyses. These two occasions have different purposes and therefore different processes. Up until this point, I have been presenting preregistration that should occur before data collection. The purpose of this preregistration is to document the idea and plan to demonstrate the a priori nature of the research. There are many options offered for such a preregistration, though I use the COS preregistration template because it seamlessly fits into the OSF project workflow. The purpose of freezing the materials on the repository after data collection and analyses happens is to promise future researchers that the materials, data, and plans presented at the time of publication will not be altered or deleted in the future. This process is better termed registration because it happens after rather than before the project is conducted. It is generally the last step for a project after it is accepted for publication.

PREREGISTRATION AND THE UNDERGRADUATE RESEARCH PROJECT

For the students not planning on graduate school and those with the intention of graduate school that is not research focused, this chapter might feel disconnected from current goals of completing a course to finish a degree. One might ask, "What is the value of preregistration when there is no intent to publish?" The answer is that student work is always published, even if there is only one reader. Students can use preregistrations to confidently demonstrate knowledge to the instructor the same way researchers can do this with reviewers and editors. For students in my courses, there is very little extra time in the semester between the idea generation, data collection, and finishing the term. By engaging actively in the preregistration process, students are taking advantage of the front-loading of workflow. Rather than trying to write up details late in the term, when there are many other time costs, documenting these details early encourages writing when timelines are more flexible.

Similarly, because students are learning about the research process, reviewing the questions on a preregistration template helps preview researcher degrees of freedom. This can be illuminating for questions that students have not fully considered, such as how many subjects are needed or what will be done with bad data or how variables will be combined into constructs. Learning about each of these is a common part of any research methods course, and preregistration is one more avenue for students to engage and consider this material.

Chapter 3: Crisis Schmeisis Journey Book Review

The Negro Revolt

Louis Lomax

Written in the early 1960s, this book explains the challenges of race relations 50 years ago. It is striking how much has not changed even as we try to make progress in these areas. The reader might also be surprised by Lomax's predictions about how the civil rights movement would progress. His perspective is very helpful in understanding the diversity of responses in civil rights movements (https://osf.io/tz6e7/).

CHAPTER 3 EXERCISES

Exercise 3.1: Reviewing Book Project Preregistration

- Download a copy of the book project example preregistration (http://osf.io/mp7yt/).
- Review each question and critically evaluate whether you would have a different answer or what is missing.

Exercise 3.2: Completing a Power Analysis

- Download G-power—https://stats.idre.ucla.edu/other/gpower/.
- Online option: www.ai-therapy.com/psychology-statistics/power-calculator
- Choose appropriate statistics and estimate effect size.

Exercise 3.3: Preparing Methods

- Assemble materials for research (novel or crowd).
- Imagine yourself as a participant and identify all materials and procedures that need to be controlled.
- Put materials on the OSF component.
- Write paragraph describing procedure in OSF Materials component wiki.
- Videotape the procedure with a volunteer and review for ways to improve.

Exercise 3.4: Preparing a Preregistration

While the final version of a preregistration needs to be located on a time-stamped repository, Open Stats Learning and Project Tier partnered to develop a preregistration template for students to practice. In this template, answer each question with as much detail as possible.
After reading the basic introduction on the OSL/PT main Page,

- Follow the link to their OSF page.
- Download the OSL/PT Preregistration Template.
- Complete all the questions in as much detail as possible.

END-OF-CHAPTER REVIEW QUESTIONS

1. What is the purpose of preregistration?
2. What are the researcher benefits of preregistration?
3. How does preregistration fit the diversity, social justice, and sustainability lens?
4. Describe the two families used in this chapter's NPS example. Which family do you relate to more?
5. How does preregistration help with undergraduate research projects?
6. What are three different types of preregistration, and how are they different?

References

Bem, D. J. (2000). Writing an empirical article. In R. J. Sternberg (Ed.), *Guide to publishing in psychology journals* (pp. 3–16). Cambridge: Cambridge University Press.

Bernieri, F. J., Gillis, J. S., Davis, J. M., & Grahe, J. E. (1996). Dyad rapport and the accuracy of its judgment across situations: A lens model analysis. *Journal of Personality and Social Psychology*, *71*(1), 110–129. doi:10.1037/0022-3514.71.1.110

Clark, C. J., Winegard, B. M., Beardslee, J., Baumeister, R. F., & Shariff, A. F. (2020). Retracted: Declines in religiosity predict increases in violent crime—but not among countries with relatively high average IQ. *Psychological Science*, *31*(2), 170–183. doi:10.1177/0956797619897915

Flake, J. K., & Fried, E. I. (2020). Measurement schmeasurement: Questionable measurement practices and how to avoid them. *Advances in Methods and Practices in Psychological Science*, *3*(4), 456–465.

Hall, B., Wagge, J., Pfuhl, G., Stieger, S., Vergauwe, E., IJzerman, H., . . . Moreau, D. (2018). *Accelerated CREP-RRR*. Retrieved from https://scholar.google.com/citations?user=1qve7Z4AAAAJ&hl=en

Sue, D. W., Capodilupo, C. M., Nadal, K. L., & Torino, G. C. (2008). Racial microaggressions and the power to define reality. *APA PsycNet*, *63*(4), 277–279.

Turri, J., Buckwalter, W., & Blouw, P. (2015). Knowledge and luck. *Psychonomic Bulletin & Review, 22*, 378–390. doi:10.3758/s13423-014-0683-5

Wicherts, J. M., Veldkamp, C. L., Augusteijn, H. E., Bakker, M., van Aert, R. C., & van Assen, M. A. (2016). Degrees of freedom in planning, running, analyzing, and reporting psychological studies: A checklist to avoid *p*-hacking. *Frontiers in Psychology, 7*, 1832. doi:10.3389/fpsyg.2016.01832

4 DECISION HEAVYWEIGHTS

Drawing Inference With Confidence

Chapter 4 Objectives

- Learn why it is important to choose a statistical path for analyses
- Understand difference between statistical significance and significance
- Define statistical significance in relation to the book hypothesis
- Understand different types of sampling methods
- Explore how sampling introduces bias into data collection
- Learn about random assignment
- Explore frequentist, Bayesian, and estimation processes

MUSIC IN A BOX: ABOUT THE SONG "DECISION HEAVYWEIGHTS"

"Decision Heavyweights" is intended to be a clever musical allegory that features the conflict between the frequentist and Bayesian approach to inferential statistics as two opponents in a boxing ring. The conflict between the two approaches masks the real power player, which is confidence intervals (CIs) and effect sizes, which is the only evidence convincing enough to reject the null hypothesis (defeating the null named Pops). The opening lick draws a "p" or a "B" on the guitar neck, and the song travels through experimental territory with lyrics steeped in metaphor.

MAKING DECISIONS ABOUT DECISIONS

This chapter features the only song in which I employed a strong metaphor: two boxers in a ring. As a pacifist, this metaphor is particularly violent for me. I did this because these fierce debates still rage between the factions supporting one of three distinct approaches: frequentist (traditional p-values), Bayesian statistics, and effect size/CIs. Readers should not expect a correct answer from me. I do have a preference (bias) for approaches for my circumstances, particularly as it relates to working with undergraduate students. By the end of this chapter (or the song lyrics), my bias will be clear, but I do not suggest that my decision is better than those of my colleagues, who seem just as convinced that they are right.

Ideally, researchers reading this book have identified their research question, identified measures and methods, and are prepared to collect data. This chapter helps to clarify the decisions about how to determine whether hypotheses should be supported or rejected. Students will likely need to adhere to their instructors' choices, but it is valuable to make our own decisions as independent researchers, and good decisions come from critically evaluating the choices. My critical evaluation offers a starting point for others who are still thinking about it. To help conceptualize this content, the rest of this chapter continues to advance this material through the consideration of the hypothesis related to visitor age and miles walked.

RESEARCH IN ACTION: CONSIDERING STATISTICAL SIGNIFICANCE USING THE NATIONAL PARKS QUESTION

It is better to think about statistical significance conceptually before numerically. And before considering statistical significance, consider just significance. Alone, significance means important; something with weight and consequence. However, when the adverb "statistically" is added to make a two-word phrase, the meaning no longer suggests importance, it means unlikely. This needs to be repeated because the point is so significant! Statistical significance does not imply importance, only unlikeliness to have occurred by chance. When reporting outcomes, researchers should be clear about their own use of these two terms. Just because something is unlikely to have occurred by chance does not make it important. To highlight this, return to the national parks research question and consider it both from the question of whether the findings have importance and whether they likely occurred by chance.

First, what is important about the question of whether there is a positive curvilinear relationship between the age of visitors and miles hiked? Frankly, I'm not sure there is any grand importance to this question at all. It is simple to talk about, and it is easy to imagine designing a study around. I do actually believe this to be true based on personal observation and basic logic. It also allows me to show expected results using a curvilinear relationship, which is difficult to understand conceptually sometimes. The question came from a more complex one that does not translate easily into the parametric statistics. I am also much less convinced of the answer, "What percentage of women aged 49 to 55 could hike 21 miles in 10 hours carrying 10 to 15 pounds with 2,000 feet of elevation gain?" I predict that number to be between 5% and 10%, and further I predict that more women than men could complete that hike. This hypothesis is personally important as my wife and I enter this age group, but I'm still not sure it has great value for the scientific community. I might argue that public-service campaigns could be developed to increase community health with the answer, but I do not claim that it is important. I do have many critically important questions about national parks, and I would encourage you to develop your own.

While testing the question of curvilinear relationships between visitor age and miles hiked might not be significant, I am very convinced that it would be statistically significant. In other words, I am fairly certain that if we measured visitor age and miles walked in an unbiased manner, we would find the resulting correlation to be different from zero. Because correlation is not causation, the research question offers no explanation for why this might occur. However, if we find something to be statically significant, we should expect it to occur again . . . and again . . . and again. We would not expect the same numbers to result each time; we would not even expect all the correlations to be statistically significant. However, we would predict that more times than not, it would be big enough to consider it different than zero.

In inferential statistics, the probabilities associated with whether an effect is big enough to be considered real are called p-values. In psychology, at the turn of the millennium, researchers were dedicated to null hypothesis significance testing approaches to inference and typically accepted p-values of 0.05 or less as evidence that an effect was real or not. The consideration of how often p-values should be statistically significant provides some evidence of the reproducibility of an effect according to Simonsohn, Nelson, and Simmons (2014). If an effect is real, p-values should be closer to 0.000 than 1.000, and there should be many more p values < .01 than between .02 and .05. There is an exercise connected to a tutorial to help interested readers learn how to conduct a p-curve analysis for their own areas of interest. Ideally, this would help demonstrate that any scientific decision is based on probabilities, and the pattern of these statistics should follow other mathematical principles.

TESTING WITH AN UNBIASED SAMPLE FROM A KNOWN POPULATION OF PARKS

In the preregistration of the book research question in Chapter 3, I documented the number of subjects needed, the intended measures, and the intended analyses. However, without appropriately sampling, any conclusions cannot be trusted. This section is intended to both reinforce learning about sampling and highlight how sampling and statistical significance rely on each other. In the end, a statistic is an estimate of the likelihood that we will find the same thing again if measured in the same population. For this estimate to be valuable, the sample needs to be random and unbiased. This section will consider methods of reaching that unbiased sample and how the result is related to the rest of the population.

The premise of statistical significance assumes that a statistic is computed based on measurements of an unbiased sample. The best way to achieve this is through simple random sampling. For simple random samples, each and every member of the population has an equal probability of being selected. As is true in most studies, it is not possible to achieve simple random sampling in the national parks study. Try to think of a way to do that. What location makes it possible to access each person in any park? They all have multiple entrances, they are all very large, and people access them across all 24 hours in a day. Further, national parks are open 365 days a year, and people engage in different activities across the year. I hope the reader sees the futility and expense with trying to conduct a simple random sample. To address the impossibility of this situation, we have more feasible options.

Other forms of random probability sampling include stratified random sampling and cluster random sampling. Either of these would be much more feasible in this case. Stratified random sampling measures randomly from groups of the population, and cluster random sampling measures randomly from different locations. They similarly approach the problem of feasibility by segmenting the population into more manageable chunks. Here, I offer an example of each version but encourage readers to imagine other approaches that could be employed. In each example, I will offer a feasible and affordable version, assuming the appropriate collaborators and permissions. I will also suggest how that study could be set as ideal but would then only be feasible with large grant support (i.e., many millions of dollars would be needed).

Stratified Sampling

Stratifications within the population represent meaningful and mutually exclusive categories of people. In this case, we would try to identify meaningful groups and then randomly sample within those groups. For the question about age and miles walked, what meaningful categories might be useful to collectively represent the entire population? In this case, we are seeking a cost-saving device (time is money in research too). After much consternation, I can only think of one cost-saving form of stratification, "purpose of visit." All other examples do not seem to reduce costs or cannot be as easily determined during testing. If the reader identifies a better version, I invite them to email me, and I will include it in online materials. To take advantage of "visit purpose" as stratification in a feasible sample, I would invite all visitors of the park in a given period of time to answer a one-question survey about why they were coming to the park: single-day sightseeing, overnight car camping, single-day hiking, overnight hiking, passing through, other. After they answered that question, I would ask them to continue in the survey if they were also randomly selected to be sampled according to their group membership. This is feasible at peak times, because lines at entrances can be more than a mile long. I have measured such lines while driving home multiple times. At slow times, we might not have as many volunteers to stop, but ethically, we can only invite participation, not dictate it. The ideal version of this would sample from all the parks rather than one or a few.

Cluster Sampling

The procedure of cluster sampling is to divide the locations that hold the population. The parks being sampled would be divided up into regions, and then everyone in that region would be recruited for the sample. In the feasible study, the number of parks and number of locations would be more limited. These limitations would reduce generalizability of the feasible study, because the findings would only apply broadly to the parks sampled and not to all parks. In the case of the ideal example, all parks would be sampled, and then clustering would occur only in locations within parks, not across parks. Because I am most familiar with Mt. Rainier National Park, I will use this as an example. In terms of where to sample (see Figure 4.1), the following locations could be reasonably staffed to get frequent and unique visitors: (Campgrounds) Ohanapecosh, Cougar Rock, White River; (Visitor Areas) Longmire, Henry M Jackson, and Sunrise; (Popular Trails or Easy Access Destinations 1–2 mile) Grove of the Patriarch, Frozen Lake from Sunrise, Narada Falls, Sunrise picnic areas, Paradise picnic areas, Kautz Creek parking lot, Longmire old growth trail, Reflection Lakes, and Sunrise Point. We would sample randomly from this list of 15 locations depending on how many researchers could help collect data and then randomly assign a researcher to each location to go administer the survey to individuals selected to participate following established procedures that could vary from further randomization or selection guidelines.

Nonrandom Sampling Options

I wish I had the resources to randomly sample from the population. The one time I tried resulted in two different publications (Hauhart & Grahe, 2012; Grahe & Hauhart, 2013). However, most of the research I have conducted and most of the research published in psychology employed convenience sampling. The readers are likely planning to conveniently sample as well. If you are intending to seek volunteers from the university subject pool, that is convenience sampling. It is not representative of all the students at the school,

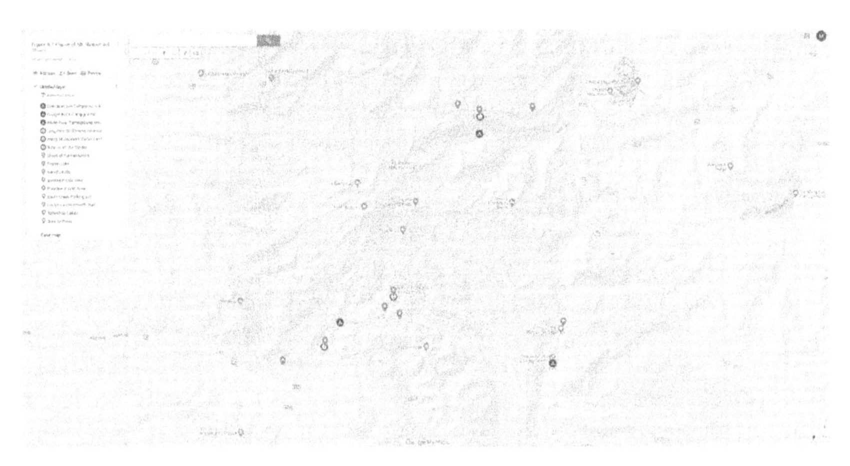

Figure 4.1 Figure of Mt. Rainier Split Into Clusters

www.google.com/maps/d/u/0/edit?mid=1VWQ2t_pMrDhB5gsjGqa87kwY7HUrZ6hv&usp=sharing

only those taking certain courses. In fact, any open recruiting seeking any volunteer to complete a study is convenience sampling.

Convenience sampling has a host of problems, and each increases the possibility of bias in the sample and therefore the statistics. To start with, volunteers are not like normal people; they are more conscientious and more agreeable (Lönnqvist et al., 2007). Another issue is that convenience samples can end up with uneven variability of the predictor variables. In the case of this example, we could conveniently sample by standing at all three visitor centers or entrances and invite anyone to complete the survey. One way to reduce the problem of bias in the convenience sample would be to purposefully sample from each age group until sufficient numbers were present. Otherwise, if just seeking anyone, the sample could end up with too many (or not enough) members of one of the age groups. Even with purposive sampling, convenience sampling of any kind violates a basic premise of inferential statistics, randomization.

RESOLVING SAMPLE BIAS WITH RANDOM ASSIGNMENT

Because most research methods are experimental, the readers should be relieved somewhat that random assignment saves convenience samples. While convenient samples violate statistical assumptions, random assignment of subjects within a convenient sample across experimental conditions resolves the violation. For this reason, psychological researchers can confidently study questions with convenient samples. However, researchers are then required to generalize their conclusions only to individuals within that sampling frame (the population from which the convenience sample was drawn).

If I added a second question regarding "motivation to hike" in the visitor age-by-miles-hiked example, I would predict a different set of outcomes across ages. Whereas I predict a curvilinear relationship between age and miles hiked in ability, I would predict a linear relationship between age and miles desired to hike. If we also measured age ordinally rather than via interval data, this study would become a 2 (question: motivation × ability) × 7 (age: child, <12; teen, 13–18; emerging, 19–25; adult, 26–35; mature 36–50; middle, 51–70; older, 70+) design. The design could be either fully between or mixed where the question variable is mixed and the age variable is within. A fully within design would require either a longitudinal study or a study using recall from people old enough to have been in all seven groups. Arguably, neither of these is feasible, so the design works best as either a between or mixed. The results would likely be different, but those statistical considerations are for a different book focused more on statistics. Also, we could use regression to measure the same questions being addressed in the factorial design, and with more precision.

In this section, the point to recognize is that this study would likely be different depending on where the researcher found the convenient sample. If collected at the national park entrances or visitor centers, this would likely still get a wide variety of ages representative of larger populations. But what if the study were posted online or conducted using your institution's subject pool?

Perhaps the park superintendent would only allow data collection in the campgrounds or in parking lots. The answers might be very different depending on where in the park the visitor is. Visitors found in the backcountry are more likely to be experienced hikers and campers who enjoy long hikes to reach those areas. On the other hand, the campgrounds provide a safe, monitored area to set up camp and access to shorter trails for those with less experience or less interest in difficult hikes. What if the study could only be conducted in the backcountry, 2 or more miles from any visitor center? The sample frame would suggest very different populations in each of these cases.

UNDERSTANDING FREQUENTIST STATISTICS THROUGH ANOVA

In order to get the most of this section, readers should refamiliarize themselves with basic statistics concepts related to central tendency, variability, normal distributions, *z*-scores, *z*-tests, and *t*-tests. In this section, I will demonstrate how an ANOVA would be approached from a frequentist statistics perspective. A brief revisit of the underlying assumptions of ANOVA can remind readers of some critical fundamentals of frequentist approaches. ANOVA assumptions include: (a) independently sampled observations, (b) samples drawn from normally distributed populations, and (c) homogeneity of variance. The purpose of ANOVA is to then measure whether variance attributed to the category variable (age) is greater than the variability attributed to error variance. The only difference between ANOVA and *t*-tests or *Z*-tests is that they are intended for more than two groups at a time. Thus, at the center of the question for ANOVA is whether the observed F is greater than the critical F that marks the edge of the area of the distribution beyond which the researcher agrees that the effect was due to the manipulation or category variable rather than sampling variability.

Until this chapter, I presented the visitor age-by-miles-hiked example as a correlation testing a curvilinear function. In fact, this is the effect that is being studied by the Able column in Table 4.1. The reader can test this effect with these estimates by squaring the average of the age category and correlating that with the able miles hiked estimates. The data can then be used to compute a correlation estimate. Is there an effect present? If the line depicting those

Table 4.1 Predicted Estimations for Number of Miles Hiked During 24 Hours

	Able	*Willing*
Child	2	1
Teen	20	0
EmAdult	24	12
Adult	34	20
Mature	30	24
MidAge	12	36
Older	4	40

means is flat, then there is no effect. If the line is sharply angled, there is a strong effect. A slight angle might be significant, but only with many subjects or much statistical power. In this case, the hypothesis is supported if the line looks like an upside-down horseshoe. The deeper the curve of the shoe, the stronger the effect. While this is computed based on seven scores in this example, we would not check significance on a correlation table. Instead, it is the r effect size, which would be computed from an ANOVA measuring maybe 200 or more respondents.

Before continuing, I insert the Null Hypothesis Significance Testing (NHST) steps associated with ANOVA. NHST is the frequentist's tool in causal inference and valuable to review (see Table 4.2).

If this were an ANOVA with $n = 10$ respondents for each condition, and we looked only at the Able condition, the question being asked is, "Are the means more variable than one would expect if they were all from the same population?" In other words, the ANOVA would test the assumption that the different age groups yield distinct populations of miles hiked. Instead of thinking about this from the perspective of means, imagine the sampling distributions that those seven groups represent. A frequentist approach determines if the variability assigned to the effect exceeded the variability assigned to error in a large enough proportion to say that it was unusual enough that chance did not cause the variability. A frequentist would check an F with a certain value between (6) and within (63) degrees of freedom against a chart (or check the p-value associated with it) and decide whether it was bigger than necessary to reject the null hypothesis. A frequentist doesn't care how big the curve of the horseshoe is or how angled the straight line is, only whether it is different from zero. The frequentist is guided by the p-value, and the decision is arbitrary and complete. An effect either exists or doesn't, there is no close. If it does, any size is big enough to consider worthy of reporting.

I will compare and contrast these at the end of this section. Astute readers might recognize that Table 4.1 offers a factorial design rather than a single one-way ANOVA. This will be considered in greater detail in Chapter 5. For now, the demonstration is only to highlight a frequentist approach to the basic

Table 4.2 Null Hypothesis Decision Steps

Step	*Description*	*Book Research Question Example*
Step 1	State Null	$M_{child} = M_{teen} = M_{emAdult} = M_{adult} = M_{mature} = M_{midage} = M_{older}$
Step 2	State Alternative	the means are not equal
Step 3	Set alpha at .05	
Step 4	Set decision Rule	if $p > .05$, reject Ho
Step 5	Compute	
Step 6	Decide and report	

question, which can be addressed by positive curvilinear correlation between age and miles worked, or a statistically significant ANOVA with the appropriate pattern of means.

A BRIEF PRIMER ON BAYESIAN STATISTICS

Unlike frequentist statistics, Bayesian statistics take into account variances caused by nonrandom sampling procedures by returning a distribution instead of a single point prediction that will more often than not underestimate the amount of variance present. Priors are our preconceived ideas about the data we will be collecting. Bayesian analysis examines new data (likelihoods) as it relates to past data (prior distributions) to find the uncertainty of p (posterior distribution). In the example presented thus far, I have predicted that middle-aged individuals would walk more than those at either extreme. Basis for this hypothesis includes personal observation of various age groups while hiking and basic logic, knowing how physical development relates to physical ability. Our prior might be a known data set (such as the number of steps a person takes in a day) or a posterior predictive distribution from a previous study on similar variables. With a posterior distribution, a researcher can predict the probability of values for an unknown data point. The largest advantages of Bayesian statistics are the use of priors and the return of a distribution rather than a single point prediction. The distribution takes variances in data into account, and it can also help one determine what *N* will be needed in the study to demonstrate the validity of our result.

There are four types of distributions in Bayesian statistics. Prior distributions (predicted distribution prior to collecting data) describe the possibility of each hypothesis being true due to chance. A sampling distribution shows where the observed data would fall conditional to the set parameters of the prior distribution (i.e., a defined range). Marginal distributions allow one to look at subsets of data separated from the entire data set. Finally, we return to posterior distributions, which include all observed data and thus become a prior in future research. The posterior is the combination of the likelihood (an estimate of the rate of success given collected data) and the priors. Using this process, researchers can reject the null hypothesis while also calculating the likelihood of the alternative hypothesis being a better fit, where frequentists only look for a difference from zero.

Bayesian decisions occur in a couple ways, but the goal is always to find the optimum answer with the lowest chance of risk. The probability of error is the chance that the null hypothesis is true. One could use priors and data to decide on the likelihood that a person of a certain age will be able to hike a number of miles. One can also turn to Bayesian statistics to perform A | B tests of which hypotheses are likely to be true. To calculate Bayesian distributions, researchers at the University of Amsterdam have created a free, open-source program known as JASP (Jeffrey's Amazing Statistics Program; https://jasp-stats.org/). JASP works on multiple platforms and can be run directly through a browser. With a user-friendly and attractive GUI (graphic user interface), researchers can load data sets and then drag and drop variables to see results. As variables are added or removed, the results on screen update in real time. While originally developed to handle Bayesian calculations, frequentist statistical tests have also been included, making it possible to compare the results between the two approaches. There are several sample data sets already loaded into the program so that those new to the concept can practice before collecting their own data.

ESTIMATION APPROACHES WITH EFFECT SIZES AND CONFIDENCE INTERVALS

Researchers from both frequentist and Bayesian approaches make decisions based on a probability value that a statistic occurred by chance. How they approach those statistics and which distributions they use to check those probabilities are slightly different and might yield different conclusions, but they still have the same ultimate reliance on a p-value. In contrast, effect size approaches focus on other conditions to draw conclusions. Estimators are more likely to focus on whether the expected effect is big enough to make an impact or the estimated differences are large enough to confidently accept. To understand effect sizes, it is helpful to consider visualizations of effects as represented on Figure 4.2 with the 95% "cat's eye" distributions drawn around the subsample means. Figures 4.2a and b were created using R 4.0.3 (R Core Team, 2021), along with the psych (v2.0.9; Revelle, 2015), multicon (v1.6, Sherman, 2015), readxl (v1.3.1; Wickham & Bryan, 2019), and tidyr (v1.1.2, Wickham, 2020) packages.

To start, review the 95% CI of miles hiked regardless of age based on my estimated data [12.23,16.57]. If these data were real, it would suggest that we are 95% confident that the true population mean was between these two points. That does not indicate that the mean does fall between those numbers as sample variability always allows for the possibility that any data set is misrepresenting reality. It also does not suggest that the population mean is at the center of those numbers (the sample mean). However, it is more likely that the population mean is nearer to the center than toward the tail. The 95% CI represents the range of scores that would be represented if the sample considered were converted into the t-distribution. Perhaps the formula is worth printing here:

> 95% CI = M + /— tcrit * SD/sqrt(n), where tcrit given df from sample @ .05, 2 tailed.

Notice that the formula is converting the M into t units by multiplying the SD times the t associated with a two-tailed alpha = .05. Confidence researchers recall that the t-distribution is just the Normal distribution converted into a more platykurtic shape given that they are drawn on samples rather than populations. Therefore, when there is an estimate that mu is between X and Y,

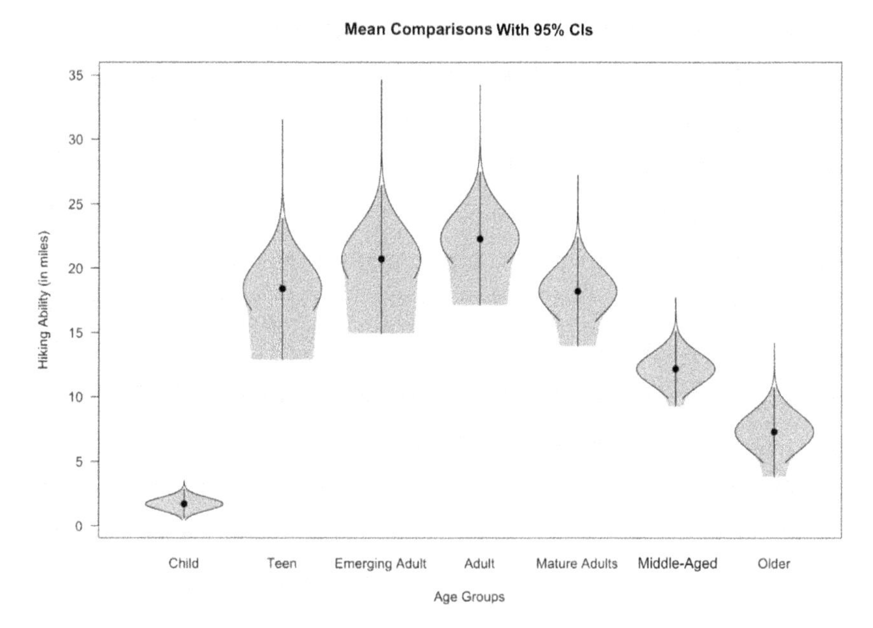

Figure 4.2A Ability Mean Comparisons With 95% CIs

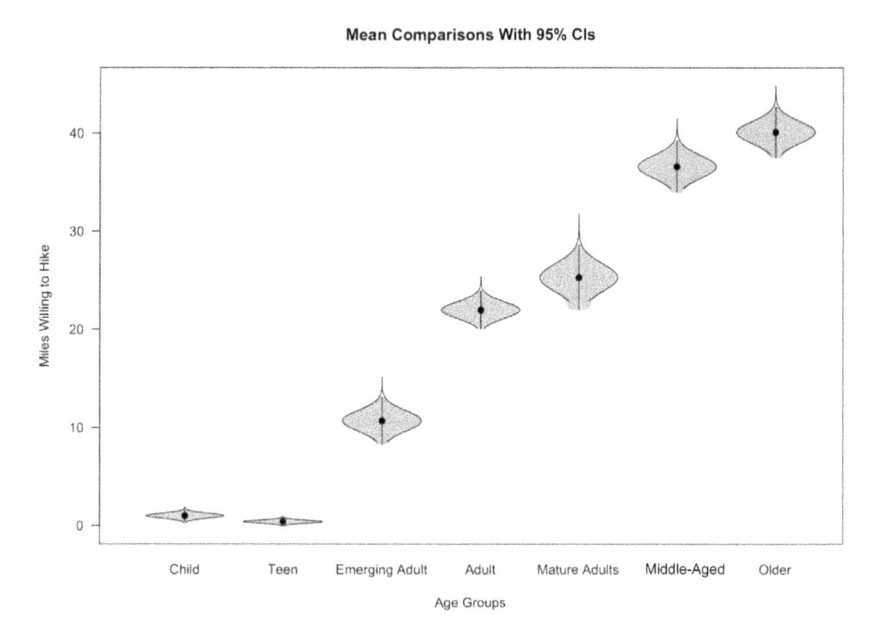

Figure 4.2B Willingness Mean Comparisons With 95% CIs

the estimate is truly reflecting the probability that it is each number in between as well. This helps when looking at a set of means and looking for patterns, as with the recurring age and miles question. When looking at seven groups that are intending to display a curvilinear effect, there should be a pattern in the means showing an upside-down horseshoe. However, the pattern is not sufficient; we want to see differences between groups. Maybe not complete separation for each group, but good distance across groups with no overlap. Given that these data were made up by me to demonstrate an effect I believe to be true, the reader should not be surprised that the data clearly demonstrate expectations. Hopefully, as researchers, we would all be so lucky to always find our expected results.

The other estimation approach is to look at effect sizes, which are statistics that reflect the magnitude of an effect. A commonly reported effect size for basic statistics is Cohen's d, which reflects the mean differences converted to standard deviation units, and effect size r, which presents the ratio of explanatory to total variance. This is then squared for eta squared and then interpreted as the proportion of total variability that is explained by the effect. Much like CIs are derived from frequentist sampling distributions, effect size r and eta squared can be viewed alongside basic inference statistics.

In the case how many miles people of different ages can hike, the simulated F (6, 63) = 16.66, $p < .001$ represented a frequentist demonstration of the effect. The associated effect size eta squared can be derived using the formula eta square = SSeffect/SStotal. An estimator would see these results and conclude that there is a strong curvilinear relationship between age and miles hiked. Remember that the frequentist drew the conclusion that there was an effect and described a pattern of means that matched that conclusion.

However, the estimator offers more information by including a size of effect. Further, the estimator can share a CI around the effect and suggest the possible range of the population effect. Table 4.3 displays the means, standard deviations, and 95% CIs for both Willing and Able simulated data. These data were simulated using the same formulas as employed in the data set associated with Exercise 4.1, which generates random numbers, but the values were saved to provide permanent statistics. This data file (https://osf.io/fqdr8/)

Table 4.3 Means, Standard Deviations, and 95% CI Limits for Ability and Willingness to Hike Example

Age group	M_{able}	SD_{able}	Low_{able}	$High_{able}$	M_{will}	SD_{will}	Low_{will}	$High_{will}$
Child	1.70	1.57	0.58	2.82	1.00	0.82	0.42	1.58
Teen	18.40	7.56	12.99	23.81	0.40	0.52	0.03	0.77
EmAdult	20.70	7.89	15.06	26.34	10.70	3/37	8.29	13.11
Adult	22.30	7.09	17.23	27.37	22.00	2.71	20.06	23.94
Mature	18.20	5.81	14.05	22.35	25.30	4.52	22.06	28.54
MidAge	12.20	3.99	9.34	15.06	36.60	3.66	33.98	39.22
Older	7.30	4.76	3.89	10.71	40.10	3.57	37.54	42.66

can be found at the Book Example Spreadsheet on the data and analysis component of the book example OSF page.

DECIDING ON DECISIONS

This chapter presented three competing approaches to analyzing data. The frequentist and Bayesian approaches both rely on hypothesis testing with p values, where the estimators use the same underlying math to estimate the range and size of effects using the same sampling distributions. APA requires researchers to present both inference tests and effect size indicators, and this offers the richest information for the research consumer. Beyond that, I encourage readers to critically evaluate their own situation when deciding between frequentist and Bayesian approaches with a strong or weak emphasis on estimation. Here are a few points to consider: purpose, audience, and existing knowledge.

The intended outcome of the research has great influence on this choice. Exploratory research is better served with estimation decisions. Conversely, confirmatory research can be advanced with tests of both the null and alternative distributions as provided by Bayesian approaches. Some journals will be more responsive to certain approaches. *Basic and Applied Social Psychology* (BASP) made quite a stir in 2015 when the editor announced there would be no more p-values published in their journal, essentially requiring estimation decisions (Trafimow & Marks, 2015). Also, the audience should be considered, because some reviewers will expect Bayesian rather than frequentist or estimation rather than others. If the author is writing in a subdiscipline that values one over the other, it would be challenging to advance in the circle without adopting the common practices. Finally, what is known about the effect? The major advantage of Bayesian over frequentist approaches is the use of priors to make a more fine-tuned estimate of the effect. If testing a frequently published effect with the goal of making strong inference, Bayesian will be more convincing.

Whichever method is selected, the question of how that effect replicates over time is important. If an effect has been tested multiple times, it is possible to use the effect sizes as data points and measure whether the collective outcome is statistically significant. This approach is called a meta-analysis. Compared to crowd projects that try to test effects with many very large samples, meta-analyses try to estimate the size of effects by collecting all the published and unpublished estimates of a given effect and converting them all to the same effect size metric. The average effect size can then be compared against zero.

While these methods can be extremely complex, Wilson (2014) offers an introduction to meta-analyses for amateurs with links to free software. Her instructions help novices identify criteria for study inclusion, how to find the effects, and a basic guide to what happens in the analyses. Though practicing meta-analyses is beyond the scope of this book, there are many guides for how to conduct them (c.f. Siddaway, Wood, & Hedges, 2019).

In the end, science is advanced by findings that are reproducible and meaningfully sized. As I note in the lyrics of this chapter's song, either stat needs to bring a big swing to beat the null named Pops (the null hypothesis). In other words, if the effect size is not big enough to replicate, then over time we will accept the null anyway.

Chapter 4: Crisis Schmeisis Journey Book Review

Warriors Don't Cry

Melba Pattillo Beals

This book is a firsthand account from one of the Little Rock Nine, who bravely faced white mobs when they attended a school they were previously banned from attending. This book resonates with the content of this chapter because everyone involved in desegregation was making a choice. The decisions were guided by different approaches to life and yielded remarkably different results from violence to prevent social justice to bravery in the face of existential threat to advance it (https://osf.io/tz6e7/).

CHAPTER 4 EXERCISES

Exercise 4.1: Estimating Simulated Effects: Book Research Question

Download the Exercise 4.1 Data and Formulas.xlsx (https://osf.io/k5y8z/) file from the data component on the Book Example OSF page.

- Review the Data Generation sheet.
  - Offers formulas to estimate 10 subjects for each of the seven age groups for both Ability (cells A1-H12) and Willingness (cells L1-S12)
  - Prewritten formulas to measure M, SD, SE, and 95% CIs for both Ability (cells V1-AD17) and Willingness (cells AH1-AP17) by copying and pasting the simulated data.
  - The M, SD and 95% CI values are also collected for both variables in a single table (cells V22:AC30) and then rotated into a usable format in (AG22:AP29).
- Review the One-Way ANOVA Sheet.
  - This includes a location to copy and paste data from previous sheet (cells B2:H12), which will automatically compute an F with a summary table (A23:F26).
  - Sources of Variance for Total SS (cells I1:P13), Within SS (cell Q1:X13), and Between SS (A14:H17)
- Review the Factorial Data-able Sheet.
  - This includes formulas to simulate a 5 (age: emerging, adult, mature, middle, older) × 3 (difficulty: easy, moderate, strenuous) between-S design data set (cells A1:G31)
  - A table of group means (B33:G36)

- Review the Factorial ANOVA Sheet.
    - This includes a location to copy and paste data from the previous sheet (cells C2:G31), which will automatically compute an F with a summary table (A39:F44).
    - Sources of Variance for Total SS (cells Q1:W33), Within SS (cells I1:O33), and Between SS (A50:H52)
    - A means table for group means (C33:G36), Marginal Means of Age (B37:G37), and Marginal Means of Difficulty (H33:H36)

After you become familiar with the simulation formulas, the following could be fun and interesting:

- Change the formulas.
    - These formulas reflect my own ideas; make them closer to your ideas.
    - Review how changes in the formulas impact the F and CIs.
- Conduct a mini meta-analysis
    - Change the random values on the data generation sheet with any keystroke.
    - Paste special (values only) and collect the data from multiple samples.

Exercise 4.2: Stating the Design

- List all manipulated variables (V1, V2, V3).
    - If zero, then not experiment
- List the number of levels for each manipulated variable (#V1, #V2, #V3).
    - If any are one, then it is a control variable, not manipulated.
- Will participants experience all levels of manipulation (Within), or will there be different participants at each level (Between) or is it both (Mixed)?
- State the design using the following formula
    - #V1 × #V2 × #V3 Between/Mixed/Within-Subjects Design
- For the Book Example
    - 7 (Age) × 3 (Difficulty) Between-Subjects

Exercise 4.3: Estimating Simulated Effects: Own Research Question

- Make a copy of the Book Research Question Excel Sheet.
- Change design to match your own.
- Alter formulae to make the scales match your measures.
- Run analyses.

- Was the effect statistically significant?
- Was effect size small/large?

Exercise 4.4: Explore the Dance of p-Values

This program shows how results of p-values should occur given certain parameters. By changing the parameters, you can explore outcomes associated with Null differences or large differences. The following are numbered as the creator of the program did. Do it multiple times and explore the dance of the p-values (www.esci-dances.thenewstatistics.com/).

Decisions

1. The population: Adjust the mu and sigma to desired numbers; choose distribution.
2. Click to Display (population, SD lines, fillrandom)
3. Controls (clear, take sample, Run/Stop)
4. Samples (N, Data points, sample means, dropping means)
5. Mean Heap (mean heap, sampling distribution curve, SE lines, +/- MoE around mu
6. CIs (95% or other; assume sigma is known or unknown)
7. Capture of mu (mu line, capture of mu)
8. Capture of next mean
9. Dance of the p values (sound, volume)

Run and then stop, change parameters, and watch again.

Exercise 4.5 Conducting a P-Curve Analysis

- Go to BiTSS P-Curve Tutorial and read material.
  - www.bitss.org/p-curve-a-tool-for-detecting-publication-bias/
- In addition to basic instruction, they identify three criteria for study p-values to be included.
  - associated with the hypothesis of interest,
  - statistically independent from other selected p values, and
  - distributed uniformly under the null.
- Go to online p-curve app: www.p-curve.com/app4/.
  - There are existing values that show a p-curve that suggests that effect is real.
  - Add new values until the p-curve no longer shows an effect.
- Conduct an actual p-curve analysis.
  - Identify criteria for effect of interest.

- Find papers reporting effects meeting criteria.
- Replace existing values at p-curve.com with values from effect of interest.

END-OF-CHAPTER REVIEW QUESTIONS

1. Describe the difference between significant and statistically significant.
2. List the three types of statistical approaches described in the chapter.
3. How does sampling create bias?
4. How do we correct for sampling bias?
5. List three points to consider when choosing a statistical approach.
6. Compare and contrast two ways of sampling a population. What are the advantages and disadvantages of each?
7. Why is it important to consider your statistical approach before collecting data?
8. Describe what effect size is and how conclusions are based off of it.
9. What are ways in which you can incorporate the ideals of diversity, social justice, and sustainability when considering sampling techniques and statistical approaches?
10. Describe the advantages and disadvantages of both the frequentist and Bayesian approaches.

References

Grahe, J. E., & Hauhart, R. C. (2013). Describing typical capstone course experiences from a national random sample. *Teaching of Psychology*, *40*(4), 281–287.

Hauhart, R. C., & Grahe, J. E. (2012). A national survey of American higher education capstone practices in sociology and psychology. *Teaching Sociology*, *40*(3), 227–241.

Lönnqvist, J. E., Paunonen, S., Verkasalo, M., Leikas, S., Tuulio-Henriksson, A., & Lönnqvist, J. (2007). Personality characteristics of research volunteers. *European Journal of Personality: Published for the European Association of Personality Psychology*, *21*(8), 1017–1030.

R Core Team. (2021). *R: A language and environment for statistical computing*. Vienna, Austria: R Foundation for Statistical Computing. Retrieved from www.R-project.org/

Revelle, W. (2015). *Psych: Procedures for personality and psychological research*. Computer Software Manual, R Package Version 2.0.9. Retrieved from https://CRAN.R-project.org/package=psych

Sherman, R. A. (2015). *Multicon: Multivariate constructs*. Computer Software Manual, R Package Version 1.6. Retrieved from https://CRAN.R-project.org/package=multicon

Siddaway, A. P., Wood, A. M., & Hedges, L. V. (2019). How to do a systematic review: A best practice guide for conducting and reporting narrative reviews, meta-analyses, and meta-syntheses. *Annual Review of Psychology*, *70*, 747–770.

Simonsohn, U., Nelson, L. D., & Simmons, J. P. (2014). P-curve: A key to the file-drawer. *Journal of Experimental Psychology: General*, *143*(2), 534.

Trafimow, D., & Marks, M. (2015). Editorial: Banning null hypothesis significance testing procedures. *Basic and Applied Social Psychology*, *37*(1), 1–2.

Wickham, H. (2020). *Tidyr: Tidy messy data*. Computer Software Manual, R Package Version 1.1.2. Retrieved from https://CRAN.R-project.org/package=tidyr

Wickham, H., & Bryan, J. (2019). *Readxl: Read excel files*. Computer Software Manual, R Package Version 1.3.1. Retrieved from https://CRAN.R-project.org/package=readxl

Wilson, L. (2014, September 30). *Introduction to meta-analysis: A guide for the novice*. Retrieved February 8, 2021, from www.psychologicalscience.org/observer/introduction-to-meta-analysis-a-guide-for-the-novice

5 ODE TO P-HACKING

Making Decisions Before They Happen

Chapter 5 Objectives

- Define researcher degrees of freedom
- Explore why replications are seen as easier than novel research
- Understand the decisions that researchers make during the research process
- Understand where researcher degrees of freedom are in each section of a paper
- Learn about how researchers detect questionable research practices

MUSIC IN A BOX: ABOUT THE SONG "ODE TO P-HACKING"

"Ode to p-Hacking" is a tongue-in-cheek lament about the days in which a researcher could go back and make their hypothesis fit the data they analyzed; a time when papers had stronger support of hypotheses and more straightforward narratives. While it led to bad science, the papers might have been more fun (and quicker) to read. These decisions, even unintentionally made, increase the likelihood of finding a $p < .05$ when the initial result was $.05 < p < .10$ or otherwise nonsignificant. Large-scale replications, preregistered hypotheses, and considering effect sizes are all methods to reduce p-hacking. The chord progression lifts heavily from Delta blues and the lyrics inspired by a Tool song, "We Knew the Pieces, They Would Fit." This song also hints at the many places in a manuscript that p-hacking can occur.

REMEMBERING HISTORY IS EASY; PREDICTING THE FUTURE IS THE OPPOSITE

When I was five years old, my family went to visit a farm in the village to which we had just moved. I remember how empty the countryside was and how stark a change it was from my toddler years in Baltimore. Going out to pet cows and accidentally walking through a manure field are such vivid memories of that day. It is a lovely story, which I can clearly recall as I type these words. But in reality, I do not know if this was my first trip to a farm or even if we saw that scenery as we drove down that road. When I visited that village for the first time 40 years later while taking my child on a college visit, so much looked strangely the same. And some memories returned that had been long forgotten as we looked at my childhood church (which was closing) and briefly played on the playground of my grade school; so little had changed. However, I had the luxury of looking at the present with no ability or motivation to check the accuracy of the memories those present images were invoking.

Driving through the countryside of my youth on the opposite coast from where I now live brought me many moments of nostalgia. Part of this chapter incorporates a nostalgic interpretation of psychological research from before the crisis brought forth in the Scientific Revolution 2.0. That will be the easy part. The other part of this chapter will help readers recognize how difficult it is to predict successful hypotheses. Together, these approaches will highlight the decisions researchers make throughout the process and why they should be made before data are collected and analyzed. Or, at the least, how they should be reported if they occur after the results are known.

CRITICIZING PAST RESEARCH IS EASIER THAN TRUSTING ANY FINDINGS

One of my favorite demonstrations of the problems we face as a field was conducted by Simmons, Nelson, and Simonsohn (2011). They presented data and results in two stages, first following the reporting expectations of the time and then at a deeper level to demonstrate how they found their effect. Using random assignment of participants to conditions, they found that listening to a popular Beatles song, "When I'm 64," caused people to be chronologically younger than those who listened to a children's song. I hope the reader recognizes both the absurdity of that claim and the impossibility of actually demonstrating it. Imagine, listening to a song making someone become younger. Not just feel younger, but become chronologically younger. Or maybe that listening to the children's song made them older. If so, I'm going to change my music playlist right now. Though what they demonstrated cannot be the truth, the data were real. They conducted the manipulations. They found those results. Yet they did things to the data that they did not convey, and they engaged in procedures during the results that they did not report. They described these activities as questionable research practices. Throughout this chapter, the issues brought up in this and other examples will be explored in more depth.

Making these decisions is not what is questionable but rather the possibility of incomplete reporting. It is inadvisable to trust the reporting of these decisions, as it is too easy for them to be misrepresented by good people without malicious intent. In fact, part of the problem is that science is populated by good people who generally trust each other. In the context of a research project, the researcher has always been human, though I did recently read a *Discover* magazine story (Orlando, 2020) about robot researchers originating hypotheses about evolutionary procedures and testing them by measuring yeast production, so we might not always be able to assume a human researcher in the future. For now, though, I'll assume the reader is human.

Therefore, it is good to recognize our flaws, to recognize human flaws, as they might relate to research.

To illustrate this point, let's consider the history of the subdiscipline of social perception within psychology. Beginning in the 1940s and '50s, the questions focused on accuracy of perceptions. However, after methodological critiques of the common methods of the times, researchers shifted to studying decision errors and heuristics. By the early 1980s, most of the research was focused on studying perceptual errors rather than accuracy. It is easy to demonstrate that people make incorrect judgments about others as a result of availability of information, anchoring of evidence, contrasting conditions, and a tendency toward confirmation bias. Each of these contributes to the development of stereotyping, prejudice, and discrimination. Really, the list of examples highlighting human error is too exhaustive to consider presently and a bit disheartening to recognize.

Researchers who wanted to study accuracy during this period had to convince reviewers that the evidence they found actually represented accurate decisions because the field was so focused on human error. Eventually, a series of novel methodologies and statistical procedures allowed researchers to demonstrate that humans are in fact somewhat accurate when making judgments and that certain types of judgment accuracy can increase with time and practice. In time, the frustration with the field's general focus on demonstrating human error rather than success emerged as the subdiscipline of positive psychology, the study of how we can improve rather than document our failures. Though not focused on social perception, positive psychology approaches questions with an assumption of success.

Personally, I am a fan of studying both sides of human judgment because all accuracy research includes an error term. I also believe in having hope for the human race. Even with all our serious flaws as humans, we have the capacity to improve. Of greatest importance, we should be vigilant in seeking evidence to corroborate someone's recall of past events or interpretation of ambiguous information, particularly when they are motivated by the outcomes.

If I have convinced you not to believe someone else's claims without evidence, I have only achieved half of my goals. I would like you to stop trusting yourself as well. Students in every class I teach learn about the elaboration likelihood model of persuasion Petty & Caccioppo (1986) because it highlights why the errors are so easy to find. In order to pay attention to the real message rather than be distracted by unrelated characteristics (status, look, flash), targets of a message must be both motivated and able to discern the true message. Both of these are rarely present in most situations, which is why politicians rarely stick to the issues when they campaign and why product commercials rarely present relevant information about the product they are selling. However, for those in the know, staying motivated to avoid bias in all decisions is critical, or the thinkers become lost.

All of us can get the wrong kind of motivation and start looking for evidence only when it supports our worldview. This is called confirmation bias. Even when we are motivated and are working very hard to perceive truth in the world, we are still likely biased in our perceptions. In the end, we should not trust others or ourselves without clear evidence. That is where preregistration comes in, and this chapter will help clarify each of these researcher degrees of freedom that need to be frozen in time before data collection begins.

WHERE RESEARCHER DEGREES OF FREEDOM RESIDE

As I mentioned in earlier chapters, Jelte Wicherts and colleagues studied this problem in depth and identified decision points within each section of research manuscripts (introduction, methods, results, discussion) that should be clarified before the data are collected. These decision points can be characterized as researcher degrees of freedom. Of course, the idea of degrees of freedom can be confusing to some, and now it is being applied in a novel setting, so a brief explanation is warranted.

In many textbooks, I think authors choose a great example when they consider a softball team roster and positions to fill, but a backpacking example is a better fit for this book. Any backpacker knows the challenge of choosing what to take on a trip where there are no stores or chances to resupply. So degrees of freedom come about when considering which items to bring and where to put those items. When loading 30 to 40 items into a pack that is difficult to lift and take off, you need to plan location carefully. When stopping, the amount of time you want to dedicate to unpacking and packing varies depending on the task or item (drink, snack, bug spray, suntan lotion, lunch, new socks, other replacement clothes, jacket, first aid kit, bear spray, pocketknife). Further, when my wife and I hike together, we have to determine who carries which items that are shared (tent, sleeping bag, stove, fuel, future meals, and trash). Then there are the personables that should only come out after camp is up for the night (comfortable shoes, hygienic items, entertainment).

Imagine putting these items into your pack. What is the item that goes on the bottom first? If we imagine 35 items, the first choice allows for 35 options, 35 degrees of freedom. When the first item is chosen (in my case, it would be my bag of clothes), we will have only 34 more decisions to make. With data with a given N, once we compute a mean, we have fixed one of the data points leaving N-1 degrees of freedom. With more complex designs for ANOVA or more predictors in a regression, the number of fixed degrees of freedom increases and reduces power.

Of course, backpacks have many pockets to allow for more compartmentalization than a t-test or two-way ANOVA. For each location, the packer has decisions to make, and these decisions could impact the hike later. Unpacking an entire pack to cook lunch is time costly and may be a very dirty experience. Putting certain items in special spots and remembering where they go makes for a more pleasant trip for all involved. When my wife and I go overnight backpacking, we pack together at least one day before the trip after assembling all those items with careful planning. Better planning leads to better hiking. The research process is very similar to that. The remainder of this chapter will consider where these degrees of freedom lie within any empirical report.

When I started teaching, I would regularly talk about the difference between researchers conducting good science and what authors needed to do to get published. The good scientist records the status of researcher degrees of freedom and honestly reports them in the manuscript. The questionable scientist reports a narrative that advances theory with casual regard for the readers' need to know the decision process. Human experiences are rarely as simple as that dichotomy, and I will remind the reader that good people do questionable things unintentionally. To help clarify these researcher degrees of freedom, I will present an example effect that would be relevant to park managers (sustainability messaging and treatment of public spaces) from both the good and questionable science perspectives.

Decisions in the Introduction

Writing is hard. In the introduction, authors are tasked with explaining to the readers what an effect is, who else has reviewed it, how their findings relate to the present question, and justification for their hypothesis related to that effect. Moreover, manuscripts that are submitted for publication usually include more complexity than a simple effect. There might be a very complicated effect or multiple effects, all requiring a cohesive and yet concise introduction. Achieving this does not occur in a single writing session and does not resolve after a first draft. In graduate school, I learned to think about this in response to the critical feedback written in red ink, "think of the reader" or "you are making the reader work too hard." All of this writing and rewriting is not related to researcher degrees of freedom. More editing of the introduction is not a questionable research behavior; it is an act of consideration for the reader. Where research degrees of freedom do emerge in the introduction (Wicherts et al. identify two), they are related to the hypotheses and how they are presented.

The good scientist clearly articulates the direction and expected magnitude of an effect of an a priori hypothesis. Alternatively, good scientists who conduct exploratory research note the nature of their hypotheses and do not present them as confirmatory research. For this example, I predict that messages focused on normative pressure will increase sustainability behaviors more than other types of persuasion (fiscal appeal, nature appeal, coercive appeal). This is a priori since no data yet exist, and the direction is clearly stated. Questionable scientists present unclear hypotheses that might have been found rather than predicted. For example: I predict that messages focused on normative pressure will impact sustainability behaviors. Here, I have presented neither direction or a clear comparison of the effect.

While I do have an a priori hypothesis, the research is exploratory. It is grounded in previous research conducted by Bob Cialdini and colleagues (Goldstein, Griskevicius, & Cialdini, 2007; Nolan, Schultz, Cialdini, Goldstein, & Griskevicius, 2008) examining how messages improved sustainability behaviors such as reusing hotel towels and reducing energy use. However, there is no prior test of this specific effect. The messages and measures will be created for this study. All these combine to suggest that the hypothesis is justified, but we are not confirming an established effect; we are exploring whether a predicted effect exists. It is important to be clear about intention, because it will affect how statistics are applied to multiple comparisons later. To summarize, in the introduction, a good researcher is challenged to (1) establish whether the research is exploratory or confirmatory and (2) state unambiguous hypotheses, including the direction of the effect.

Decisions in the Methods

While Wicherts et al. (2016) identified only two researcher degrees of freedom in the Introduction, there are 11 located in the Methods section. The only way to verify the true nature of an effect using honest statistical estimates is if all information about the data is present. Early in the open science movement, Nosek et al. (2017) encourage reviewers to add this brief statement to each manuscript review:

> I request that the authors add a statement to the paper confirming whether, for all experiments, they have reported all measures, conditions, data exclusions, and how they determined their sample sizes. The authors should, of course, add any additional text to ensure the statement is accurate. This is the standard reviewer disclosure request endorsed by the Center for Open Science [see http://osf.io/hadz3]. I include it in every review.

I don't see too many reviewers add this statement to the manuscripts in my editor's queue, but it highlights in a brief statement where researcher degrees of freedom exist in the methods and design. Now I will consider each of the 11 researcher degrees of freedom from how age and difficulty predict miles hiked to the messaging and sustainability prediction.

Decisions in Conditions and Measures

Consider the implied design from the good example hypothesis earlier. There seem to be four groups (normative appeal, fiscal appeal, nature appeal, coercive appeal) and one dependent variable. The good scientist would report exactly what was conducted. A questionable scientist would follow the poor advice I received many years ago as a student. During a research meeting in which I proposed a novel study, someone argued that I should add more DVs, because I already have them in the lab and I should measure more constructs just in case my first prediction did not work out. Many questionable practices in this section are associated with this advice: (a) creating multiple manipulated independent variables and conditions; (b) measuring additional variables that can later be selected as covariates, independent variables, mediators, or moderators; (c) measuring the same dependent variable in several alternative ways; (e) measuring additional constructs that could potentially act as primary outcomes, and (f) measuring additional variables that enable later exclusion of participants from the analyses (e.g., awareness or manipulation checks).

Rather than consider each of these in sequence, recall the good scientists' approach to this section; they report in the Methods all conditions and measures. The status of the study (confirmatory or exploratory) guides the appropriateness of some other decisions, but in all cases, the most critical point is to report what was done. In a confirmatory study, the methods should be executed with constructs previously demonstrated to be reliable and valid. An exploratory study might test validity and reliability with constructs the researchers built themselves. There might be many different types of each appeal condition or other appeal messaging. The researcher might be exploring which constructs are most sensitive to the effect or be curious about how various individual differences contribute to the effect. The problem arises when the questionable scientist conducts a study such as this but then presents it as confirmatory or presents only the conditions that worked. In contrast to the four appeal conditions (normative, fiscal, environmental, coercive), a questionable scientist might report only the conditions that worked. A good scientist not only explains the exploratory nature of the study but offers all the materials, data, and analytic code so other researchers can reproduce their results.

Decision Associated With Data Collection

Two questionable research practices can occur related to data collection: failing to use random assignment or not blinding participants and/or experimenters. Both of these are directly related to the ability to draw inference in an experiment. The reason that experiments, but not observational studies, allow for causal inference is because participants are randomly assigned to a condition. This is the basic tenet of experimentation.

A brief personal experience might highlight how a questionable researcher could violate this assumption. Many years ago, my students and I created an online study where we offered four scenarios (called a, b, c, d for this example) about two people talking to each other. Because of resource limitations, rather than randomly assigning participants to a condition, we asked participants to select one of four characters (<, >, /, \) and tied one condition to each.

We assumed this would be random enough because their decision was unrelated to the study. However, when we analyzed the data, the majority of people (over 50%) selected just one of those symbols: ">." We had unintentionally created a confound variable where symbol preference (<, >, /, \) was systematically associated with condition (a, b, c, d). Any differences in the dependent variables could be due to symbol preference or condition; we had no way to distinguish. A questionable researcher might report these findings anyway and fail to specify that they were not randomly assigned. Without random assignment, causation cannot be inferred.

Like random assignment, blinded designs are a basic assumption for good experimentation. If participants are aware of specific parts of an experiment, their behavior is no longer true to their self but rather reflects the self that they wish to present. Whether the change in behavior benefits the researcher depends on how many participants respond in a socially desirable fashion rather than a contrary or reactionary fashion. All samples are biased, and though random assignment distributes that bias in an equal fashion across conditions, when demand characteristics (conditions of the experiment that inform the participant about desired outcomes) are activated, the benefits of random assignment are lost, and all conclusions are suspect.

When participants are unaware of the conditions, this is called a single-blinded design, but strong inference testing requires double-blind designs in which the experimenter is also unaware of what conditions participants experience. Experimenter bias was first documented by Robert Rosenthal (Rosenthal, 1968). After documenting the effect in a basic human research situation, he also demonstrated the effect with animal research, because critics argued that it was still the participant that mattered (Rosenthal & Fode, 1963; Rosenthal & Rubin, 1978). Interestingly, rat behavior changed due to experimenter knowledge—a much more compelling argument. He went further to document expectancy effects of authority figures by studying the impacts on children's education when kindergarten students who were randomly labeled as likely to blossom performed above their peers on class work after months of teachers demonstrating preferential treatment (Rosenthal & Jacobson, 1968). Questionable researchers might fail to note the absence of blinding.

Decisions Associated With the Analysis Plan

These practices are in the Methods portion because they should have occurred before the data collection took place. While some of these details are often reported in the results, they are associated with planning and executing the study or preparing the data for analyses, so they belong here. In this section, I will describe the questionable practices around: (a) conducting a power analysis, (b) establishing a sampling plan, (c) determining stopping rules for data collection, and (d) coding procedures.

A power analysis is necessary because if the study is conducted and null results are found, the researcher wants to guess whether there is truly no effect or if the conclusion is a Type II error (failing to find an effect because the study is underpowered). Many undergraduates conduct research for their courses and collect as much data as possible in the short period available. The question of sufficient sample size necessary for adequate statistical power is rarely considered because the number of participants is limited by time and available subjects. However, when conducting confirmatory research, researchers should conduct a power analysis to estimate the necessary sample size before data collection begins. For confirmatory studies, prior research should offer effect size estimates that can be used in a program such as G-Power to determine the minimum *N*. For exploratory studies, prior research might not be similar enough to offer a reliable estimate, and the researcher then must guess whether it is small, medium, or large. As an added consideration, the various meta-science projects (RP:P, Many Labs 1,2,3,4,5) all suggest that published effect sizes from original studies are inflated. Personally, I recommend underestimating your expected effect size to make sure you are truly powered.

When an adequate power analysis is conducted before a study begins, researchers collect data from a specified number of subjects before stopping. A questionable research practice that is related to this is determining the data collection stopping rule on the basis of desired results or intermediate significance testing. Imagine the situation in which you have a great research idea. You collect as much data as possible in one academic term and analyze the results. After conducting statistical analyses, the outcome yields an almost significant result (let's say $p = .07$). Maybe if you just had a few more subjects, the results would be significant. After collecting some additional data, the outcome reveals a significant result ($p < .05$). The problem is that the second p-value is not accurate because you violated the assumption of independence. This practice of testing and collecting until findings come out in a preferred manner elevates the Type I error rate. A questionable researcher, however, will only report the final outcome without admitting to the iterations of stopping and testing.

Related to this are methods of correcting, coding, or discarding data. Again, the research proposal should specify data transformation and coding techniques, particularly in the case of confirmatory research. However, in that situation in which the outcome is almost significant, researchers might find that removing certain respondents who yielded bad data would change the outcome in a more favorable direction. A good researcher will operationally define bad data before they analyze it by considering how long it took to complete, whether they accurately completed attention checks, or if their data represent outliers. A questionable researcher operationalizes these decisions during data collection in a nonblinded manner.

The final related practice here is associated with specifying the sampling plan and identifying the framework for testing the hypotheses. A good researcher completes a power analysis and specified coding and decision criteria a priori. A questionable researcher might collect data for running (multiple) small studies. If the questionable researcher has the goal to tell a good story about the data, using poor data cleaning technique and stopping rules based on preferred outcomes in a series of studies might look more impressive than a single study, even if sufficiently powered.

Decisions in the Results

Good scientists collect all their data and then complete a set of analyses that follow a predetermined protocol. While this is especially true for confirmatory studies, it should also be true in the first examination of exploratory data. Serendipity is a hallmark of science, such as Pavlov's discovery of the principles of classical conditioning while studying digestion. However, serendipity should follow rigor. If a study is conducted, there should be intentional analyses planned for the data, even if they are exploratory. After the planned analyses, the data can be explored inexhaustibly to find other effects. Wicherts et al. identify 15 different analysis related to questionable research practices: (1) choosing between different options of dealing with incomplete or missing data on ad hoc grounds; (2) specifying preprocessing of data (e.g., cleaning, normalization, smoothing, motion correction) in an ad hoc manner; (3) deciding how to deal with violations of statistical assumptions in an ad hoc manner; (4) deciding on how to deal with outliers in an ad hoc manner; (5) selecting the dependent variable out of several alternative measures of the same construct; (6) trying out different ways to score the chosen primary dependent variable; (7) selecting another construct as the primary outcome; (8) selecting independent variables out of a set of manipulated independent variables; (9) operationalizing manipulated independent variables in different ways (e.g., by discarding or combining levels of factors); (10) choosing to include different measured variables as covariates, independent variables, mediators, or moderators; (11) operationalizing nonmanipulated independent variables in different ways; (12) using alternative inclusion and exclusion criteria for selecting participants in analyses; (13) choosing between different statistical models; (14) choosing the estimation method, software package, and computation of standard errors; and (15) choosing inference criteria (e.g., Bayes factors, alpha level, sidedness of the test, corrections for multiple testing).

You likely noticed that some of these have overlap with those presented in the Methods section. Whereas Methods includes the reporting of what conditions and measures were employed, the Results presents how the data resulting from these conditions and measures were analyzed. Rather than considering each of these 15 practices uniquely, I offer an example of a project that could inspire both quality and questionable research.

Take, for instance, the Emerging Adulthood Measuring Multiple Institutions (EAMMI) study (Reifman & Grahe, 2016). This crowdsourced survey was planned and executed by Alan Reifman in 2004 to test the relationship between experiences of emerging adulthood and people's political decisions and ideas. With more than 1,300 respondents from 10 locations, he tested his exploratory hypotheses and found no support. As a good scientist, he stopped using the data and turned his efforts in another direction. However, because he invited contributors to add scales, the resulting survey included scales that allowed for incidental findings because they were not previously tested together.

In 2016, the *Emerging Adulthood* journal published nine empirical reports analyzing different questions in this data set. Given the success of the EAMMI in yielding publishable data, Grahe et al. (2018) developed the Emerging Adulthood Measuring Multiple Institutions 2: The Next Generation (EAMMi2; http://osf.io/te54b/) as a conceptual replication. These data have so far resulted in seven published studies, and though the EAMMI data are not public, the EAMMi2 data are publicly available. This is a good moment to invite readers to see if they can find other meaningful effects that need to be published out of these data. For researchers approaching these data, those 15 analysis practices need to be used by good scientists, but they become questionable when they are poorly reported or reported in a misleading fashion.

Decisions in the Discussion Section

The final set of decisions is presented in the discussion section along with potential supplemental materials associated with the projects. Wicherts et al. (2016) state that these six questionable research practices are associated with reporting. To avoid two of these, (a) failing to assure reproducibility (verifying the data collection and data analysis), and (b) failing to enable replication (rerunning of the study), researchers should offer all their materials, data, and analysis plans as supplemental material. Doing this when publishing with some journals will result in receiving Open Science Badges to mark the transparency of the researcher. In Chapter 3, I presented preregistration as an open science tool. However, Wicherts et al. found some who failed to mention, misrepresented, or misidentified the study preregistration in their final report.

Failing to share a preregistration because it is counter to the outcomes is similar to another practice, failing to report so-called "failed studies" that were originally deemed relevant to the research question. Researchers who are pursuing a research question might plan 3, 4, 5, or more studies to address multiple questions related to a core effect. Even if some of those studies do not demonstrate the effect, they should all be reported, because the meta-analytic effect size should be calculated with all data, not just data that supports the outcome. Earlier, I mentioned that metascience research has regularly estimated smaller effect sizes than the originally published study. One explanation for this is that the published research only included data or studies that supported the conclusions, but a good scientist reports it all.

Some research practices are not decisions that are ever appropriate; they are always wrong. Misreporting results and p-values to augment supporting arguments violates basic ethics in publishing. Mistakes may happen as we copy numbers from one place to another, but intentionally presenting findings as stronger than they are is not a mistake. Sacha Epskamp and Michelle Nuijten developed a program called Statcheck (http://statcheck.io/) to help authors and editors catch honest errors. The program quickly assesses the reported statistics, degrees of freedom, and p-values to determine if they are all consistent. If not, the author can figure out their error and fix it before publication.

The final questionable research practice is connected to many of the earlier ones described: presenting exploratory analyses as confirmatory. Throughout this chapter, I have repeated that many of these decisions are not inappropriate if they are honestly presented. In each case, presenting something as confirmatory when it was exploratory is problematic. Kerr (1998) described this class of behaviors as hypothesizing after the results are known (HARKing). This process violates basic statistical assumptions and misrepresents all findings. It is tempting to HARK when strong findings come out in the opposite direction from what was intended. A good scientist would need to explain the original justification in the introduction, and a new explanation for both the effect and why the effect did not support the expectations. A questionable researcher who is HARKing only needs to present the justification that is consistent with the results. It is a shorter paper with less conflict—a much easier

Table 5.1 Comparing Good vs. Questionable Scientists

Good Scientist	*Questionable Scientist*
Records and reports researcher degrees of freedom honestly	Reports a narrative that advances the theory rather than explains details
Clearly articulates direction and magnitude of hypothesized effect	Presents unclear hypotheses that have been found, not predicted
Reports methods, measures, and analyses exactly	Collects data from running studies until desired results occur
Explains any exploratory nature of the study	Presents exploratory work as confirmatory
Presents all materials, data, and analytic code	Reports only the conditions that worked

paper to write. In Table 5.1, the good and questionable science practices are listed side by side. Consider all the ways that these behaviors could manifest in your own research if you didn't protect yourself with preregistered plan.

DETECTING QUESTIONABLE RESEARCH

Before focusing on methods to find bad research, it is valuable to remember that good people can conduct questionable research, as we sometimes mislead ourselves. Many initiatives in the open science movement are developed to prevent errors from happening. However, published research provides evidence for theories and future research, so we are compelled to question the veracity of known effects. When considering whether an effect actually exists, Simonsohn (2015) captures the situation metaphorically with his imagery of small telescopes being used to detect effects. Here is a colorful explanation. If someone claimed they found a new planet in the solar system, people would want to verify its existence. If the telescope they used to find it was very small, one would assume that the planet was either very big or the astronomers were very lucky. If various others tried to find the planet with bigger telescopes and they all failed, then the community would start to question whether the claim was an error. In the arena of visual magnification, the impact of size is beyond dispute; bigger telescopes see objects with better clarity than smaller ones do. Simonsohn (2015) argues the same should be true in the science of psychology. And though humans are more variable than planet orbits, the preponderance of evidence over time should point to effects being true. If many highly powered replications fail to discover an effect that a researcher previously demonstrated with an effect with low power, one would start questioning whether the effect was there at all. And so the crowd projects are one form of that massive replication effort. But other tools are also offered as methods to determine the reproducibility of effects.

Reviews of responses to the replication crisis (Renkewitz & Keiner, 2019; Shrout & Rodgers, 2018) discuss both procedural and statistical tools to address the problems. Many of the procedural steps echo content presented in earlier chapters, like open data/materials/process or projects that take advantage of larger samples like crowd projects. Other suggestions focus on changes to the review process or journal expectations, which are presented in later chapters. Collectively, the statistical procedures might be broadly categorized into the following: (a) procedures testing statistical assumptions, (b) conducting unbiased meta-analyses, and (c) employing Bayesian approaches.

Of note, Renkewitz and Keiner (2019) also explain detailed instructions and offers supplemental examples for statistically ambitious readers. One example of procedures testing statistical assumptions was the p-curve analysis presented in Chapter 4, where Bayesian approaches were also introduced. Another example of a procedure testing statistical assumptions is the use of funnel plots, where the y-axis tracks effects reliability (using standard error) and the x-axis represents the effect size. Evidence of publication bias is suggested when there are many data points outside the funnel. This would suggest that studies with small samples identified effects more effectively than those with large samples.

Meta-analyses are mentioned repeatedly in this text. In this context, the challenge is avoiding making errors interpreting poorly constructed meta-analyses. As with other types of research, decisions made during the process (such as about study selection and effect coding) can impact statistics and therefore interpretations. Care should be taken to document the process of meta-analyses as with any other research to allow others to reproduce the analyses and challenge the conclusions.

CONCLUSIONS

Wicherts et al. (2016) identified 32 different questionable research practices that can occur. Though there is overlap, they describe some in each section of a research paper. A good scientist avoids questionable research practices by planning in detail in advance, following that plan, and sharing their materials and data. A questionable scientist manipulates and measures broadly in hopes of finding something that can be justified later and keeps the materials and data private to avoid being proven wrong. The replication crisis occurred due in part because writing good research papers is not as easy, and the world they present is not parsimonious. Conducting good science requires ongoing vigilance so that the entire experiment can be known in order to draw strong inference related to the findings.

CHAPTER 5: CRISIS SCHMEISIS JOURNEY BOOK REVIEW

The Last Indian War: The Nez Perce Story

Elliot West

This book recounts the background and outcomes of the wars with the Nez Perce. In a chapter devoted to questionable research practices, this book reminds readers that movies and television tell stories that are more interesting even if they are not true. It is hard to read this book without feeling grief for the treatment and misrepresentation of these people, particularly since there were so many other horrendous acts toward Native Americans before that (https://osf.io/tz6e7/).

CHAPTER 5 EXERCISES

Exercise 5.1: Finalize Project Registration

- Using the OSF, complete the registration that was drafted for Chapter 3.

Exercise 5.2: Estimating Limitations: Connect Research Decisions to Research Outcomes

- Imagine that your results are all nonsignificant. All of them.
- What conclusions would you draw about the effect or theory?
- What factors might indicate that it was a Type II error (sample size, random error)?
- What would convince you it was a real finding and the effect was nonexistent?
- Repeat steps, but assume effect was statistically significant.
- What factors might indicate that it was a Type I error?
- What would convince you it was a real finding?
- Was the amount of evidence different for Type I and Type II errors?
- Save this work for the discussion section after results are known.

Exercise 5.3: Identify Questionable Decisions

Make an expanded table of research decisions and describe how a "good researcher" and a "questionable researcher" would respond to your own research project. This exercise is intended to help students avoid letting decisions wait till later.

END-OF-CHAPTER REVIEW QUESTIONS

1. What song reportedly caused people to decrease age chronologically?
2. Define researcher degrees of freedom.
3. Explain the difference between "good" and "questionable" researchers.
4. What are the two degrees of freedom in the introduction?
5. When is the best time to make decisions about the research process?
6. Describe questionable research practices and explain how it can be easy for a researcher to engage in those practices unintentionally.
7. Explain researcher degrees of freedom and list five areas in which it is important to consider them.
8. How does the consideration of degrees of freedom stay in line with the ideals of diversity, social justice, and sustainability?

9. What are some ways in which scientists can detect questionable research?
10. What is p-hacking?

References

Goldstein, N. J., Griskevicius, V., & Cialdini, R. B. (2007). Invoking social norms: A social psychology perspective on improving hotels' linen-reuse programs. *Cornell Hotel and Restaurant Administration Quarterly*, *48*(2), 145–150.

Grahe, J. E., Chalk, H. M., Alvarez, L. D. C., Faas, C. S., Hermann, A. D., & McFall, J. P. (2018). Emerging adulthood measured at multiple institutions 2: The data. *Journal of Open Psychology Data*, *6*(1).

Kerr, N. L. (1998). HARKing: Hypothesizing after the results are known. *Personality and Social Psychology Review*, *2*(3), 196–217.

Nolan, J. M., Schultz, P. W., Cialdini, R. B., Goldstein, N. J., & Griskevicius, V. (2008). Normative social influence is underdetected. *Personality and Social Psychology Bulletin*, *34*(7), 913–923.

Nosek, B. A., Simonsohn, U., Moore, D. A., Nelson, L. D., Simmons, J. P., Sallans, A., & LeBel, E. P. (2017, June 12). *Standard reviewer statement for disclosure of sample, conditions, measures, and exclusions*. Retrieved from osf.io/hadz3

Orlando, A. (2020, May 19). *Will A.I. make medicine more human?* Retrieved January 27, 2021, from www.discovermagazine.com/technology/will-a-i-make-medicine-more-human

Petty, R. E., & Cacioppo, J. T. (1986). The elaboration likelihood model of persuasion. In *Communication and persuasion* (pp. 1–24). Springer, New York, NY.

Reifman, A., & Grahe, J. E. (2016). Introduction to the special issue of emerging adulthood. *Emerging Adulthood*, 135–141.

Renkewitz, F., & Keiner, M. (2019). How to detect publication bias in psychological research. *Zeitschrift für Psychologie*, *227*(4), 261–279. doi:10.1027/2151-2604/a000386

Rosenthal, R. (1968). Experimenter expectancy and the reassuring nature of the null hypothesis decision procedure. *Psychological Bulletin*, *70*(6, Pt.2), 30–47. doi:10.1037/h0026727

Rosenthal, R., & Fode, K. L. (1963). The effect of experimenter bias on the performance of the albino rat. *Behavioral Science*, *8*(3), 183–189.

Rosenthal, R., & Jacobson, L. (1968). Pygmalion in the classroom. *The Urban Review*, *3*(1), 16–20.

Rosenthal, R., & Rubin, D. B. (1978). Interpersonal expectancy effects: The first 345 studies. *Behavioral and Brain Sciences*, *1*(3), 377–386.

Shrout, P. E., & Rodgers, J. L. (2018). Psychology, science, and knowledge construction: Broadening perspectives from the replication crisis. *Annual Review of Psychology*, *69*, 487–510.

Simmons, J. P., Nelson, L. D., & Simonsohn, U. (2011). False-positive psychology: Undisclosed flexibility in data collection and analysis allows presenting anything as significant. *Psychological Science*, 22(11), 1359–1366.

Simonsohn, U. (2015). Small telescopes: Detectability and the evaluation of replication results. *Psychological Science*, *26*(5), 559–569.

Wicherts, J. M., Veldkamp, C. L., Augusteijn, H. E., Bakker, M., van Aert, R. C., & van Assen, M. A. (2016). Degrees of freedom in planning, running, analyzing, and reporting psychological studies: A checklist to avoid p-hacking. *Frontiers in Psychology*, *7*, 1832. doi:10.3389/fpsyg.2016.01832

6 YOU CAN'T PLAGIARIZE YOURSELF

Avoiding Errors With Ethical Writing

Chapter 6 Objectives

- Discuss making ethical choices everywhere, with a focus on national parks
- Define ethics more fully in relation to professionalism
- Discuss ethics in open science, TOP, and COPE guidelines
- Further distill the idea of ethics particular to a psychology researcher
- Look at general principles and APA guidelines on writing
- General recommendations on writing

MUSIC IN A BOX: ABOUT THE SONG "YOU CAN'T PLAGIARIZE YOURSELF"

This song was written in response to a student's query in 2004 about whether they could use their work from another class in their research methods paper. The answer is, "You can't plagiarize yourself" with two meanings. First, if you own the copyright, it isn't plagiarism to reuse your own material. Second, if you did publish it, such as use it for another assignment, you shouldn't reuse it without permission. This song was not originally slated as part of this album, but the Scientific Revolution offers tools that identify academic dishonesty. After a major case of self-plagiarism emerged, this song became again timely and still informative. For students, the song allows for discussion about how to avoid academic dishonesty for their own work. Follow these lyrics to writing with integrity. This fun, fast-paced rocking song invites the audience to join in the chorus.

RESEARCH IN ACTION: BEING ETHICAL EVERYWHERE

National parks cannot operate without trust. Even if fully funded, the parks cover hundreds of square miles with minimal personnel. Rules to protect the wondrous resources cannot be enforced with authoritative pressure, though threats of violations carry federal rather than state or local punishments. Yet laws, rules, and regulations that govern us require trusting visitors who engage in backcountry hiking or camping. One can walk for miles on marked trails without seeing another person, and the chances of solitude increase should someone leave the trails (in itself a possible rule violation). In short, park authorities are asking visitors to obey laws with little to no oversight. Some of these laws, rules, and regulations are clearer and more universal to follow than others. For instance, murder, assault, and theft laws apply here like everywhere else. However, in a national park, theft includes the animals (poaching), plants, and rocks as well as items and objects. Other rules that confuse people include forbidding pets to be on trails, let alone letting them be off-leash. For that matter, in some areas, humans are not even supposed to be off-trail. In the fragile tundra and fields, human steps can cause damage to rare and threatened species. In campgrounds, visitors are not supposed to collect wood for their fires. I have seen many rule violations in national parks. When possible, I try to educate an offender. However, in the end, they will act as they choose, and I have little power to stop them beyond words and maybe a stern glare.

This helps us understand the ethical quandary of researchers, because the scientific community must trust the community to be good. Much like hiking in the backcountry, scientists can engage in much of their work in solitude. And much like the rules and regulations governing human behavior in national parks or in the world around us, some are clearer and more universal to follow than others. For instance, fabricating data is something that is clearly wrong, and a large survey of psychology researchers suggests that it is very rare (< 2%). John, Loewenstein, and Prelec (2012) included two conditions of their survey, with one using a validated method to increase truthful answers. When not motivated to be truthful, only 0.6% of respondents admitted to faking data, but only 1.7% reported that they had faked data in the truthful condition.

Much like violent acts and major crimes, a small portion of the population will violate clear rules. However, John et al. found much higher levels of questionable ethical behaviors for other behaviors that were more ambiguous, particularly at the beginning of the 21st century. For instance, more than 50% of the sample admitted to failing to report all measures, changing their stopping rules after checking results, and selectively reporting studies that worked. Other self-admitted behaviors that occurred for 20% to 50% of the sample included: failing to report all conditions, stopping data collection early because results were found, rounding a p-value down, excluding data after looking at the data, and reporting unexpected results as being predicted from the start (HARKing). In fact, of 10 behaviors, only 2 occurred at low numbers (making questionable claims about the data was below 5%). Clearly, when left to their own devices, psychology researchers were not adhering to the science equivalent of "staying on the trail." Thus, this chapter considers the ethics of research in greater detail.

This book generally helps new researchers present their work in transparent ways. This chapter focuses on the ethical issues associated with writing and reporting that research. It is beyond the scope of this book or chapter to address broader ethics in research, such as how to complete and administer informed consent. It is also beyond my expertise to argue about the complexity of ethical decisions across distinct moral frameworks. For instance, I will not consider the long-term consequences of basing decisions on values such as universalism and self-determination versus security, tradition, and conformity. Rather, this chapter focuses on the ethics of the reporting and publication process. However, I do hope that the reader intends to display ethical behavior in their lives every day, everywhere they exist. Being ethical requires one to determine personal values, identify life decisions and outcomes that support those values, and then enact behaviors to facilitate those decisions and outcomes. Aligning goals and behaviors ethically can be challenging, as sometimes it is easier to resolve inconsistency with subpar outcomes, as highlighted by the psychological phenomenon of cognitive dissonance. To that end, I will briefly share my own challenges with aligning goals and behaviors to encourage your own personal journey.

Elsewhere, I introduced the importance of valuing diversity, social justice, and sustainability in psychology (Grahe, 2019). My own preference toward self-determination, universalism, and hedonism (from Schwartz's basic human values, 1992, 2012) was instilled by my family and life circumstances. It likely brought me to the profession of college professor, because the tasks that I do to succeed in this job align with those values, in my view. However, I was also raised in a time and place that did not recognize systematic racism. Though I was regularly socially liberal compared to my peers, I was not sufficiently educated to recognize how my own stereotypes and prejudices blinded me to others' experiences. I did not see my white male privilege, nor did I fully recognize how society removed privilege from others. It was not until graduate school that I began to understand these broader issues.

Two decades later, when a PLU self-study found that the community collectively valued diversity, social justice, and sustainability; I recognized a new framework that could help guide me ethically. I have pursued a deeper understanding of how valuing these can advance a better humanity since 2012, and during my sabbatical, I dedicated a major portion of my time to self-reflection and reading, including the books that I selected along the way to encourage deep reflections about diversity, social justice, and sustainability.

While some readers might find these values cumbersome or challenging to enact personally, it is the journey of finding and enacting values that forms the basis for ethical decisions. In the present context, I will note that after extensive study, my colleagues and I presented the argument that open science initiatives promote diversity, social justice, and sustainability Grahe, Cuccolo, Leighton, and Alvarez, (2020). As we progress through the remaining ethical considerations, the reader can assume this preference will guide my own actions.

BEING ETHICAL PROFESSIONALLY

Colleges and universities are bastions of knowledge, and education is the process of transferring that knowledge from one generation to the next. Ideally, this process would occur without fault, always approached ethically. However, colleges and universities are institutions built by people, and the process requires people for both the delivery and the reception of knowledge. Unfortunately, it is common knowledge that wherever there are people, ethics falter. To help guide humans through the ethics of education, statements of ethical expectations are codified by institutions and shared through publicly located "academic integrity" statements such as those on provosts' webpages and course syllabi. The definition located on my favorite open access educational resource, Wikipedia, describes the most commonly understood principles. "Academic integrity supports the enactment of educational values through behaviors such as the avoidance of cheating, plagiarism, and contract cheating, as well as the maintenance of academic standards; honesty and rigor in research and academic publishing."

Researchers should avoid all of these same behaviors, though human researchers have demonstrated poor behavior in the past. Of greater interest to this chapter is the final subclause, "maintenance of academic standards; honesty and rigor in research and academic publishing." While students in most classes are not concerned with this portion of academic integrity statements, it is a major aspect of research ethics.

Ethics and Open Science

Chapter 5 introduced a series of decisions made during the research process that are collectively referenced as researcher degrees of freedom. When the researcher is vague or sloppy or fails to provide transparent evidence, they are engaging in questionable research practices. When the researcher is either misleading or fraudulent in reporting how these decisions were made, these questionable research practices are essentially cheating the process. One benefit of scientific transparency is that it makes it easier to avoid cheating. My own experiences with students cheating suggests to me that most instances of cheating are acts of confusion or desperation rather than preplanned and intentional. I imagine that most scientists who also fall into cheating activities do so from a place of confusion or desperation. Remember that the "publish or perish" mentality is a real threat to many livelihoods. Rather than valuing the transfer of knowledge, the academic system places great weight on success, with major penalties for failure. This creates an environment conducive to cheating (Herndon, 2016; Grimes, Bauch, & Ioannidis, 2018). While there are sad examples of academic cheating by professionals, I prefer to assume most of us are trying to be good. However, even the ideal researcher is a flawed human, as I introduced in Chapter 1. We need help to maintain not only ethical but fully transparent research.

The *Transparency and Openness Promotion (TOP) Guidelines* offer clear recommendations to help that flawed human achieve ideal research. These TOP guidelines were developed collaboratively between publishers, grant funders, and researchers to collectively represent how research should be transparent. The TOP guidelines include eight categories with four possible levels (see

Table 6.1). The eight categories (citation standards, data transparency, analytic methods transparency, research materials transparency, design and analysis transparency, preregistration of studies, preregistration of analysis plans, and replication) reflect all aspects of research conducted across any discipline. The levels represent degrees of compliance such that in Level 0, the journal might encourage the practice; Level 1, the author declares the status of the practice; Level 2, the journal requires the author to follow the practice; and Level 3, the journal verifies the information provided by the author.

When first developed, journals were invited to become signatories where they declared that they would do a self-audit and determine the level of transparency that they wanted to achieve. As the managing executive editor

Table 6.1 TOP guidelines with four possible levels

Guideline	*Not Implemented*	*Level 1*	*Level 2*	*Level 3*
Data Transparency	Journal encourages data sharing or says nothing.	Article states whether data are available, and, if so, where to access them.	Data must be posted to a trusted repository. Exceptions must be identified at submission.	Data posted to a trusted repository; reported analyses will be reproduced independently prepublication.
Research Materials Transparency	Journal encourages materials sharing or says nothing.	Article states whether materials are available, and, if so, where to access them.	Materials posted to a trusted repository. Exceptions must be identified at submission.	Materials posted to a trusted repository, and reported analyses will be reproduced independently prepublication.
Study Preregistration	Journal says nothing.	Article states whether preregistration of study exists, and, if so, where to access it.	Article states whether preregistration exists and allows journal access during peer review for verification.	Journal requires preregistration of studies and provides link and badge in article to meeting requirements.
Replication	Journal discourages submission of replication studies or says nothing.	Journal encourages submission of replication studies.	Journal encourages submission of replication studies and conducts results-blind review.	Journal uses Registered Reports as a submission option for replication studies prior to observing the study outcomes.

at *The Journal of Social Psychology*, I encouraged the board to become signatories. Already, we were encouraging open science by awarding badges (Grahe, 2014), but the process helped us decide to slowly advance our required levels of transparency. As we implemented processes to move us firmly into Level 1 across most categories, we also decided to move to Level 2 by first requiring authors to achieve research transparency (Grahe, 2018) and then data transparency (Grahe, 2021).

My own ethics do not require Level 3 from any of these categories because readers become the verification system of published work meeting Level 2 criteria. In other words, if researchers share data and materials in a public repository, any reader can verify or challenge the work presented in the paper. The researchers are therefore motivated to be honest to protect their future reputations. Researchers who engage in too much academic dishonesty can lose the ability to publish in their field or lose their job. Thus, if the researchers are willing to share their data and materials, I am willing to trust that they put forth their best effort in presenting their research.

Authors might only interface with the TOP guidelines indirectly as they prepare their research for publication. For instance, when we enacted the materials and data transparency requirement, most authors in the first few months did not recognize the instructions as they prepared their work. I created a document called "transition" in which I collected statements that I would send to authors to direct them either to the instructions or to how to implement them. Our journal was the first in my subdiscipline to require materials and data transparency, so authors needed to be trained. Naturally, so did reviewers and editors as we all experienced a new format of presenting our work.

Freedom of the Academic Press

Committee on Publication Ethics (COPE) Publication Guidelines are located at https://publicationethics.org/files/u7141/1999pdf13.pdf. *The Journal of Social Psychology* could not advance scientific transparency without approval from our publisher, Taylor and Francis. Ultimately, they decide what gets published. Without the COPE publication guidelines, publishers might be confused about that decision process. Much like the First Amendment of the US Constitution guarantees free speech and a free press, the COPE publication guidelines state that publishers should maintain that same freedom of the press in academic and research publishing.

As a private, international company, Taylor and Francis is not required to follow US laws in all its publishing spaces. That does not mean that a company is engaged in illegal activity, but not all countries adhere to freedom of the press in their laws. So I was heartened to learn about the COPE guidelines. These guidelines attempt to establish ethics across all areas of the research process with 10 broad categories (study design and approval, data analysis, authorship, peer review, conflicts of interest, redundant publication, plagiarism, duties of editors, media relations, advertising). See Table 6.2 for descriptions.

Table 6.2 Sample of COPE Guidelines

Category	*Action 1*	*Action 2*	*Action 3*
Study design and ethical approval	Laboratory and clinical research should be driven by protocol; pilot studies should have a written rationale.	Research protocols should seek to answer specific questions rather than just collect data.	Protocols must be carefully agreed upon by all contributors and collaborators, including, if appropriate, the participants.
Data Analysis	All sources and methods used to obtain and analyze data should be fully disclosed; detailed explanations should be provided for any exclusions.	Methods of analysis must be explained in detail and referenced if they are not in common use.	The post hoc analysis of subgroups is acceptable as long as this is disclosed.
Peer Review	Suggestions from authors as to who might act as reviewers are often useful, but there should be no obligation on editors to use those suggested.	The duty of confidentiality in the assessment of a manuscript must be maintained by expert reviewers, and this extends to reviewers' colleagues, who may be asked (with the editor's permission) to give opinions on specific sections.	The submitted manuscript should not be retained or copied.
Duties of Editors	Editors' decisions to accept or reject a paper for publication should be based only on the paper's importance, originality, and clarity and the study's relevance.	Studies that challenge previous work published in the journal should be given an especially sympathetic hearing.	Studies reporting negative results should not be excluded.

In contrast to the TOP guidelines, which focus, on the type of information that should be shared, the COPE guidelines offer specific statements to encourage ethical behavior and reporting. There is much that overlaps as the reporting standards for transparency coincide with ethical behaviors. The categories related to data collection and analyses direct the researcher in activities related to how the planning and administration of the project are ethical (like role clarity and author order). However, the COPE guidelines speak to reviewers and editors in addition to the authors. The peer review process should be clear (is it blinded?), and reviewers and editors should not try to identify authors or use the data in any way. Consider that a peer reviewer receives a valuable commodity in the form of intellectual currency. The peer, as another expert, could steal that information and use it for their own purposes. Editors have more information about the author (they are never blind to author names). In their positions of power, editors need to make decisions based on the actual work and not personal opinions about the researcher or because the findings are contrary to their own position.

The COPE guidelines also speak to publication activity (redundant publication, plagiarism, and media relations) that the author might engage in but that are not directly related to scientific transparency. Redundant publication (publishing the same data or paper multiple times) is easier to catch in a modern digital age. It can be plagiarism if a publisher owns a copyright for a prior publication, but otherwise, an author being redundant is not stealing the ideas from others. The other category considering media relations makes recommendations about how to share data with mass media and how journalists and researchers should interact (e.g., researchers should know if journals are at a conference). Readers seeking to conduct and publish ethical research should only publish in journals from publishers that are COPE signatories and should critically decide how to demonstrate compliance to the TOP guidelines.

ETHICS FOR PSYCHOLOGY RESEARCHERS: AMERICAN PSYCHOLOGICAL ASSOCIATION (APA) ETHICAL GUIDELINES

The TOP and COPE guidelines apply to any research area, but those of us who work with human beings need to be more careful with our research. Chemists might need to worry about what their research could do to others if exposed to the environment, but the chemicals themselves do not require extra ethical considerations. Because I am a social psychologist trained in the US, the ethical guidelines which govern me professionally come from the American Psychological Association. The guidelines regarding human treatment were originally derived by modifying the Nuremburg codes, which emerged after the horrific science experiments conducted during World War II by Nazi doctors on unwilling prisoners as part of the Holocaust. Because these guidelines include guidance on every aspect of a psychologist's professional domain, some do not apply to us unless we are working in a treatment capacity.

However, understanding the general principles and the guidelines directly related to research and publication can help us avoid poor behavior.

General Principles

These general principles of the APA guidelines provide the underlying framework for all the sections. As I review these, I want to call them basic principles as much as general. For caring, ethical human beings, all five seem to be primary to good treatment of each other (beneficence and nonmaleficence, fidelity and responsibility, integrity, justice, and respect for people's rights and dignity). In very simple terms, beneficence and nonmaleficence can be conceptualized as helping others when we can while avoiding conflicts of interest and minimizing any potential harm. Fidelity and responsibility can be conceptualized as maintaining strong professional standards while upholding ethical guidelines. Integrity can be conceptualized as being honest and truthful. Justice can be conceptualized as expecting us to be fair and equitable in our treatment of others and avoiding letting others be unjust either. Respect for people's rights and dignity can be conceptualized as meaning that we should apply this standard to all people, regardless of their physical, historical, religious, or belief categories. In the field of psychology, we are expected to be aware of our biases and avoid letting them affect our behavior; we are expected to treat others well and to avoid causing harm through injury or conflict; we are expected to be honest and just and to treat all people with respect and consideration. The APA uses these general principles to guide recommendations in 10 sections (Resolving Ethical Issues, Competence, Human Relations, Privacy and Confidentiality, Advertising and Other Public Statements, Record Keeping and Fees, Education and Training, Research and Publication, Assessment, and Therapy). Most of these apply to professional situations beyond research, so I will only consider Section 8: Research and Publication ethics.

Section 8: Research and Publication

There are 14 ethical standards in this section to address issues at various stages of the process. Again, many of these overlap with the TOP and COPE guidelines, and this section will only briefly describe them. The 14 standards can be grouped into five broader categories related to the following issues (approval, participants, application, analyses, and publication). Approval standards (8.01. Institutional Approval; 8.02. Informed Consent for Research; and 8.03. Informed Consent for Recording Voices and Images) deal with getting permission either from the institutional review board (IRB) or from the participants for their participation. Participant standards (8.04. Subordinate Research Participants; 8.05. Dispensing with Informed Consent; and 8.06. Offering Inducements for Research Participation) address the manner in which participant approval is offered and received. Whereas 8.02 and 8.03 dictate the content of informed consent, 8.04, 8.05, and 8.06 recommend how the participant is approached with the consent request and what is offered in exchange for participation. Application standards for humans (8.07. Deception; and 8.08. Debriefing) address issues that arise during the procedure and what is shared with the participant upon completion. The application standard for animals (8.09. Human Care and Use of Animal Research) addresses all research with nonhuman live animals. The analyses standards (8.10. Reporting Research Results; 8.13. Duplicate Publication of Data and 8.14. Sharing Research Data for Verification) all address either the degree to which information about the data should be shared or sharing guidelines of the data set itself. Finally, publication standards (8.11. Plagiarism; and 8.12. Publication Credit) address who should be given credit for ideas that are published.

While each of the sets of guidelines addresses unique audience and purpose, when comparing all three (TOP, COPE, and APA), some stark similarities emerge. In all three, data and analyses should be shared whenever possible; plagiarism and false reporting are discouraged. A basic understanding of one's professional expectation and guidelines can provide a strong buffer against mistakes that could be viewed as unethical.

General Recommendations on Writing

I struggled to learn to write well as a college student. I never struggled with how to cite properly; I learned that, at least, but I wrote poorly. I failed to recognize that at the time because I was young and arrogant, but I saved some of my writing from college and early in graduate school. Reading those examples after 30 years provided a clearer perspective, and my limitations are clearer. I would like to think that some of the problems (typing errors, poor word choice) would be eliminated today with the assistance of the automatic tools in most writing software. However, my writing skills took many years to develop and still have room for growth. Improving your own writing skills will benefit both you and society, and so this section includes some basic advice I have collected and shared with my students. This generalized writing advice will help researchers maintain academic integrity and avoid ethical errors, because effortful writing contradicts accidental plagiarism. Moreover, more attention to transparency in research should help in this same regard, because documenting all aspects of the research process will reduce ambiguity and again contradicts accidental plagiarism.

Writers seeking more formalized advice can find a wealth of resources at Purdue University's Online Writing Lab (OWL, https://owl.purdue.edu/), which is the premier open-access tutorial on general writing, which additionally offers specific guidance for various disciplinary writing styles (such as APA style). Another helpful resource is offered by Marianne Fallon (2018), who offers clear and direct advice on how to write exceptional quantitative manuscripts. Her goal is to help students write APA-style papers beyond the basic content that is inserted into each section.

I offer the following bulleted list of writing tips for ease of review later. These are collected from my own experiences learning to write with instructors and peers and from working with authors as an editor.

Grahe's General Writing Advice for Research Reports

- *Write like you speak, with editing.* Students often try to write like they imagine doctorate researchers write. Instead, use your own voice when you write the first draft. Later, you should edit to remove casual language and lazy writing.
- *The best writing examples are the papers you cited.* When reading the papers that serve as sources, read them more than once. These are examples of what is good enough to get published. Try to emulate their levels of organization, support, and detail.
- *APA style prefers paraphrasing to quotations.* While other writing styles encourage many direct quotations within manuscripts, APA style does not. Paraphrasing, the author's restatement of the original work, is preferred.
- *Know what you know and what you learned.* Download copies of sources so that you can review them later. Start your Reference section as you start your paper.
- *Write what is in your paper; avoid accidental plagiarism.* The easiest way to accidentally plagiarize is to copy and paste sections from an article with the intention of editing them into paraphrasing later. To avoid accidental plagiarism, do not paste sections of published work in your paper.
- *The first draft is never good enough.* Make sure you return to your writing and edit. Always plan to complete a first draft multiple days before the due date to allow for editing time.
- *Edit after reading out loud.* When reading your own work out loud, you will notice awkward phrases or errors that you would otherwise miss.
- *In later drafts, edit from back to front.* Each paragraph should be self-contained, and reading from the back to front will help identify flaws in rationale and logic.
- *In later drafts, edit sentences from last to first within paragraphs.* Each paragraph should be self-contained, and the final sentence should flow from earlier sentences.
- *Keep the narrative moving forward.* Each section, as well as the overarching paper, contains a narrative with a beginning, middle, and end. Consider each section separately and then together.
- *Write for your audience.* The readers want to know about this topic; they are not the instructor of the course.
- *Do not expect too much from the reader.* Readers do not want to work hard at figuring out what you were trying to do; be clear, organized, and concise.
- *Use a repository to augment content.* With data, materials, analyses, and output on an OSF project or some similar repository, seamlessly direct the reader to supplemental materials.

- *Keep the narrative organized.* Your Introduction and Methods sections should anticipate the Results and Discussion sections.
- *Organize sections according to hypotheses.* Each section should follow the same basic organizational outline, following the order of the predictions or questions.
- *Keep sections separate.* There is specific content intended for each APA section. Don't mix that content between the Introduction, Methods, Results, and Discussion.
- *Use section headers.* Follow APA style guidelines and use multiple levels within sections to make it easy to find parts of interest.
- *Avoid passive voice in Methods.* Passive voice occurs when the subject of the sentence is put into the object position. APA style recommends that authors avoid passive voice throughout a manuscript. While that can be remarkably difficult in the Results and Discussion section, authors should implement active voice completely in the Methods. For example, rather than, "Participants were given informed consent by the experimenters," use, "Participants received informed consent from the experimenters."
- *Avoid redundancy.* Don't repeat statements within a paragraph. Don't repeat arguments across paragraphs. Exceptions can occur in Methods and Results, but even there, try to keep the writing interesting.
- *Review.* Read your paper as if you were a reviewer; try to remove your own bias.
- *Revise.* Don't be afraid to delete, edit, and rewrite.
- *Repeat.* Papers should be edited many times for best outcomes.

CONCLUSIONS

This chapter introduced a series of ethical guidelines that are intended to improve scientific research and reporting. The TOP guidelines offer advice on how to publish research transparently. The COPE guidelines set out ethical advice for each step of the research process and all parties involved (authors, editors, reviewers, publishers). Because I am a social psychology researcher working in the US, I follow the APA guidelines, but there are other ethical frameworks for other contexts, and it is important to know yours. Finally, I offered some basic writing advice in hopes of shortening the time it takes the reader to become a researcher with strong, publishable writing skills.

CHAPTER 6: CRISIS SCHMEISIS JOURNEY BOOK REVIEW

Breaking the Silence

Sara Alderman Murphy

This book describes the activities of the Women's Committee to Reopen Schools that formed in Little Rock, Arkansas, after the governor closed schools to avoid desegregation. This is a collection of thoughts and notes of one of the leaders of that group. When I read the book, I was shocked at how little of this history was known broadly (https://osf.io/tz6e7/).

CHAPTER 6 EXERCISES

Exercise 6.1: Ethical Boundaries: Risks at National Parks Versus in Research

- Make a list of behaviors that are against the rules in national parks.
    - What increases likelihood of compliance with avoiding behaviors?
    - What increases the likelihood of bad behavior?
- Make a list of behaviors that are against the rules in research.
    - What increases likelihood of compliance with avoiding behaviors?
    - What increases the likelihood of bad behavior?

Exercise 6.2: Evaluating Ethics: Case Study of Tuskegee Airmen Study

- Read the summary of the Tuskegee Airmen Study.
    - www.history.com/news/the-infamous-40-year-tuskegee-study
    - www.cdc.gov/tuskegee/timeline.htm
    - https://en.wikipedia.org/wiki/Tuskegee_Syphilis_Study
- What APA ethical principles were violated in this study?
- What components of systematic racism enabled these violations?
- How does systematic racism still impact research ethics today?

Exercise 6.3: Ethics of Data Simulation

Exercise 4.3 invited readers to simulate data to explore concepts related to statistical significance and effect sizes. With some more practice and time, unscrupulous people could use data simulators to create data sets that could be used unethically as fake data. Now that you have some practice with simulated data, consider the ethics of the situation.

- What open science initiatives discourage or eliminate fake data?
- What messages could be used to discourage the unethical behavior of faking data?

Exercise 6.4: Writing Checklist: Review Ways to Improve Writing While Maintaining Academic Integrity

- Open the Psi Chi APA Style Checklist: https://cdn.ymaws.com/www.psichi.org/resource/resmgr/pdfs/APAStyleManuscriptChecklist.pdf

- Complete each section as best you can.
  - Formatting your article
  - Common spelling/grammar mistakes
  - What to include in your manuscript
  - Statistics and numbers
  - Citation and references

END-OF-CHAPTER REVIEW QUESTIONS

1. What are ways in which you can be a trustworthy scientist when working on publications, posters, or presentations?
2. What are ways that open science can help researchers stay committed to ethical practices?
3. How are diversity, justice, and sustainability related to ethical research practices?
4. What are the similarities between TOP guidelines and the DJS lens?
5. Name three TOP guidelines and how you would ensure that those guidelines are followed in a hypothetical research project.
6. Where can you go to receive writing help and tutorials for APA guidelines?
7. List three of Grahe's recommendations on writing and how they can help you improve your own writing skills.

References

Fallon, M. (2018). Writing quantitative empirical manuscripts with rigor and flair (yes, it's possible). *Psi Chi Journal of Psychological Research, 23*(3), 184–198.

Grahe, J. E. (2014). Announcing open science badges and reaching for the sky. *The Journal of Social Psychology, 154*, 1–3.

Grahe, J. E. (2018). *Another step towards scientific transparency: Requiring research materials for publication*. Retrieved from https://www.tandfonline.com/doi/full/10.1080/00224545.2018.1416272

Grahe, J. E., Cuccolo, K., Leighton, D. C., & Cramblet Alvarez, L. D. (2020). Open science promotes diverse, just, and sustainable research and educational outcomes. *Psychology Learning & Teaching, 19*(1), 5–20.

Grahe, J. E. (2021). *The necessity of data transparency to publish*. Retrieved from https://www.tandfonline.com/doi/full/10.1080/00224545.2020.1847950

Grimes, D. R., Bauch, C. T., & Ioannidis, J. P. (2018). Modelling science trustworthiness under publish or perish pressure. *Royal Society Open Science, 5*(1), 171511.

Herndon, N. C. (2016). *Research fraud and the publish or perish world of academia*. Retrieved from https://www.tandfonline.com/doi/full/10.1080/1046669X.2016.1186469

John, L. K., Loewenstein, G., & Prelec, D. (2012). Measuring the prevalence of questionable research practices with incentives for truth telling. *Psychological Science, 23*(5), 524–532.

Schwartz, S. H. (1992). Universals in the content and structure of values: Theoretical advances and empirical tests in 20 countries. In *Advances in experimental social psychology* (Vol. 25, pp. 1–65). New York: Academic Press.

Schwartz, S. H. (2012). An overview of the Schwartz theory of basic values. *Online Readings in Psychology and Culture*, 2(1), 2307–0919.

7 BECOMING A SECOND STRINGER

Why Good People Do Replication Science

Chapter 7 Objectives

- Learn how to connect research projects to a variety of careers
- Explore the status of the job market
- Explain why degrees and research experiences matter
- Discuss ways to improve undergraduate research
- Look at what is wrong with traditional research methods courses
- Examine how CUREs can help and how they are better with a crowd

MUSIC IN A BOX: ABOUT THE SONG "BECOMING A SECOND STRINGER"

The seventh song of the album is a moody ballad that conveys both the anger and the sadness of young researchers. This song is a direct response to a criticism that replication scientists are just a bunch of second stringers chasing effects; this song explains why someone might be engaged in metascience or replication science and not be a second-rate researcher. The song also highlights issues of social class and privilege associated with academic outcomes. This is the only song in which different band members sing a verse to show the diversity of crowdsourcing science researchers.

CONNECTING RESEARCH PROJECTS TO CAREERS

Chapters 1 through 6 provide the basic principles that should be followed to conduct successful open science research. This chapter extends the context of research and scientific transparency to the job market and careers. In the realm of academics, the process for getting a job is distinct in that the organization is offering a contract for an extended period. Often, there is an intention that this person could receive a permanent (tenured) contract after demonstrating their job performance for some period of time. Because of the way colleges are valued in national rankings, institutions often prefer to hire people who demonstrate strong publishing and grant-writing potential. Moreover, once hired, that person is expected to demonstrate a certain level of productivity to earn tenure. This emphasis on research can cause problems and challenges in the classroom that might arise when more than a third of someone's job is to lecture and they are a good researcher/writer but a poor speaker. I hope that the reader has not experienced a professor who has many publications of articles and/or books but is not interested in capturing students' attention in the classroom. It can make a term feel unbearable.

This situation also creates an environment commonly referred to as "publish or perish," because the professor might lose their job if they do not publish enough. Both at the point of hiring and during the review process, a graduate student needs to have a strong publication record to get a job, and college professors are expected to publish as much as possible in order to get tenured and promoted. This has the effect of encouraging research that is likely to yield statistical significance. Many in the open science movement attribute incidents of questionable research practices with researcher degrees of freedom (Nosek, Spies, & Motyl, 2012). Similarly, there are recommendations for how these decisions, and thus the situation, could be improved by adopting different research and professional valuations. Personally, my commitment to crowdsourcing originated with the hope that faculty could report the work they did with students in their annual reporting (Grahe et al., 2012). In fact, the CREP, the EAMMi2, and the NICE all were developed with this intention that students, faculty, and science could benefit from the research that occurs while students learn the process.

However, while some time must be devoted to the "publish or perish" phenomenon as it impacts academia; most undergraduate students will not work in academic settings. Rather, it is important to consider the value of the research for all researchers beyond the primary task of completing the study. For students, these benefits include general life preparation for individuals not interested in pursuing extended education degrees. They also include career-specific skills that directly improve the success of students who do pursue graduate school, even if they have no research aspirations. Of great importance, underprivileged and/or disadvantaged students benefit as much or more than their peers from undergraduate research experiences but are less likely to complete them (Bangera & Brownell, 2014). Finally, for individuals committed to academia and research careers, it is valuable to understand how research experiences influence the job selection and hiring process. Before examining the intersection between research and careers, I will first more generally characterize the job market for people with psychology and related degrees.

STATUS OF THE JOB MARKET

When I was a young child, my father expected me to earn a Ph.D. He did not care what the field might be, only that his children would someday complete that degree. Later, I learned it was because he had to stop his own pursuit when my mother died during my childhood. In any case, his approach worked, because I created an expectation to become the first member of my extended family to earn a Ph.D., and I was the fifth-oldest grandchild. With this goal in my mind, I never considered any career other than that of college professor. When I talk to peers and watch students go through this same self-discovery journey, I find my experiences to be rare. In fact, earning a Ph.D. in psychology is rare (< 5 % of bachelor's students earn a Ph.D., according to the APA Center of Workforce studies, 2021). Of those who earn a Ph.D., teaching is a prominent profession (> 50%), but it is still only one option. When I look back on my own career trajectory, I see avoidable mistakes from my lack of understanding of the diversity of jobs available for a Ph.D. I find that some students have the same myopia about their possible futures as I did when I was in their place. I hope the reader has better resources than I did, but here are some thoughts and resources to consider in your own journey.

Degrees Matter, as Does Passion

It is important when considering averages to remind ourselves that the distributions often contain considerable overlap. For instance, the APA Center for Career Study reports that people with Bachelor's earn on average $48,000 a year, master's average $60,000 a year, and those with professional degrees or doctorates average $85,000 a year. However, the top ranges of some salaries for bachelors are much higher than $85,000 a year. Before I explain that in more detail, the point is still worth making that higher degrees are associated with increased income when considering all jobs and all people. To that end, I'll add that the same qualities that are associated with success in the business world are associated with success in higher education. Motivation and perseverance dictate success more than innate intelligence in graduate school.

The lesson is to find the right degree. For psychology majors, for most social science majors, the breadth of opportunity is so vast that I cannot sufficiently describe it. A psychology major might earn a doctorate in psychology (19 subdisciplines listed at the APA Careers page). However, psychology is a wonderful entry into many affiliated fields assuming the student meets program prerequisites. Law and medicine are two areas of professional degrees. But students might earn Ph.D.s in human resources, business, criminal justice, library science, social work, nursing, and so on. Beyond doctorates, there are many master's programs that allow people to help others while offering steady job opportunities, such as a masters in family therapy or masters in social work. Those who enjoy the research process, especially if they can demonstrate strong writing skills for grants, can work towards a masters in library science. Students who explore their career options before committing will be able to find a graduate program that matches their vocation most closely.

I list the vastness of graduate opportunities and assert that motivation matters to invite more people to consider something they might have not before. However, my friend, Dr. R. Eric Landrum, has repeated across his career that we need to remember that most psychology majors do not go to graduate school. He lists more than 70 jobs that BA students are qualified to hold (www.apa.org/ed/precollege/psn/2018/01/bachelors-degree), including jobs such as human resource advisor, corrections officer, real estate agent, and victim's advocate (Landrum, 2018). He notes that some of these jobs do not even require a bachelor's degree, but they have good salaries (such as a real estate agent).

What is important to remember is that the psychology major imbues students with skills that employers want. So while a real estate agent does not need a psychology degree to get the job, the skills gained in the major are associated with success in the job. Landrum and Harrold (2003) found these to be the top 10 skills expected of new employees in careers: listening skills, ability to work with others on a team, getting along with others, desire and ability to learn, willingness to learn new important skills, focus on customers or clients, interpersonal relationship skills, adaptability to changing situations, ability to suggest solutions to problems, and problem-solving skills. Many of these are skills that might occur with any major (e.g., ability to adapt to changing situations), though others might be emphasized more in the psychology major (e.g., listening to others). Fleischman, Conroy, Christidis, and Lin (2019) examined lists of job ads for psychology-related jobs and found that the most important skills for jobs requiring a high degree of psychology skills and knowledge are active listening and speaking, critical thinking, and reading comprehension. In the end, an undergraduate degree is a valuable tool to advance careers for individuals who were motivated to learn during their education.

Degrees also predict the type of job sector individuals find work in. Compiled in Table 7.1 are the 2017 estimates of employment characteristics of individuals at different educational levels. This information is compiled from the APA workforce data tools (www.apa.org/workforce/data-tools/careers-psychology). There are a number of valuable conclusions to draw from this table.

First, the likelihood of being employed is pretty good, even for bachelor's degrees, but it increases with higher education. Additionally, the work that is completed is more likely to be considered closely related to the field of psychology the higher the degree someone has earned. Simultaneously, the percentage of job categories (out of 129) decreases with higher education degrees.

Table 7.1 Career Characteristics Across Degrees

Career Category	*Undergraduate*	*Masters*	*Doctorate*
Total with degree (2017)	1,910,800	625,000	227,800
Percent Employed	72%	76%	82%
Closely Related	27%	70%	85%
Percent Job Categories	71%	57%	47%
Professional Service	2	1	1
Teaching	4	3	2
Research	6	6	3
Management	1	2	4
Sales and Marketing	3	4	6
Other	5	5	5

Similarly, a curious pattern rank ordering of job domains highlights the types of jobs that occur with different degrees. While all three degrees types include "professional services" as a prominent job domain (second or first), the other domains suggests that individuals with master's, and even more so doctorate, degrees are more likely to teach or conduct research, whereas individuals with bachelor's are more likely to be in management or sales/ marketing.

I often phrase these conclusions when I talk to students and their parents about their job future by saying, "You will find employment with a psychology degree, but if you want to do psychology, you will likely need some sort of graduate degree." There is plenty of work for someone who only holds an undergraduate degree, and they can earn as much as or more than the average employee with a masters or doctorate. The real question is, what does the person want to do?

Parks in Action: Careers With the National Park Service

Maybe the reader is interested in a job with the National Park Service. Among the many jobs beyond park ranger that are needed include at least two jobs that are social science related, data analyst and educator. And park ranger jobs require a four-year degree as a precursor to qualifying, including science and education degrees. While environmental science, biology, or chemistry might be required in some locations, psychology is a qualifying degree. For the reader who is not planning on an academic or research career, perhaps this alternate path is appealing. Or perhaps the reader would prefer to only associate time at parks with vacation and relaxation and think about all the different careers visitors possess.

Research Experiences Matter, and More Is Better

Though the focus in this chapter so far has been primarily on the psychology major, a more direct relationship can be drawn between research experiences and career success. Undergraduate research experiences vary broadly in type and offering, ranging from small studies conducted as part of a learning exercise to thesis projects led and primarily authored by students. Organizations such as the Association of American Colleges and Universities (AAC&U) describe research experiences as "high-impact" practices because they provide so many learning benefits for students (Karukstis, 2010). The evidence of this has been demonstrated through qualitative interviews (Seymour, Hunter, Laursen, & DeAntoni, 2004) as well as quantitative assessments using surveys (Bauer & Bennett, 2003; Lopatto, 2004; Taraban & Logue, 2012) or by tracking criteria across 10 years' time (Hernandez, Woodcock, Estrada, & Schultz, 2018).

Consistently, these studies find benefits in how the students identify as scientists (Bauer & Bennett, 2003; Seymour et al., 2004; Lopatto, 2004) as well as the skills that are gained (Lopatto, 2004; Hernandez et al., 2018). Of critical importance, however, is that these benefits are not consistent across discipline and demographic groups. For instance, Taraban and Logue (2012) find increased benefits for males compared to females, though Lopatto found that women were more likely to persist in science if they had research experiences. Also, Bangera and Brownell (2014) found that traditionally disadvantaged groups are less likely to be offered these experiences but that whites, compared to other groups, might be less likely to benefit from them (Russell, Hancock, & McCullough, 2007). Critically, more experience is better (Hernandez et al., 2018; Taraban & Logue, 2012), and this is not related to demographic characteristics (Hernandez et al., 2018; Lopatto, 2004). These findings support the argument that research experiences are high-impact learning experiences, and they teach skills that are desirable to future employers. Thus, even for students not planning on graduate school, there are great benefits to staying motivated and interested in research experiences.

This usefulness applies also to students heading to graduate school for degrees that are not research focused. Even for professional degrees that rarely or ever lead to research jobs (e.g., master's in social work, medical degrees, human resources), the ability to critically evaluate primary or secondary research findings as they relate to the workplace is necessary. Similarly, the project-related skills (such as time management, working in groups, written and oral communication) are directly related to project skills necessary in any professional setting.

Finally, students' capacity to conduct and complete research helps demonstrate to graduate programs that they are persistent and reliable because they can complete the research cycle. Of course, any student planning on earning any Ph.D. should seek as many research experiences as possible, since their graduate career will rely on some degree of research capacity regardless of their discipline. In short, just in case it slipped your mind, right now is a good time to send a thank-you note to your instructor for offering you an opportunity to complete research.

CHANGING UNDERGRADUATE RESEARCH

Being old enough to watch systems in education change is rewarding but also a bit daunting. College does not seem that long ago, but three decades separate me from my research methods course. I will note that many still teach the course as I took it back then. As we learned about variables, designs, reliability, and validity, we were expected to conduct an experiment to test a hypothesis. I did something with memory; I cannot recall what the question was or if the results came out as predicted. I do remember that we did not get IRB approval, I delivered the protocol in dorm rooms and classroom spaces, and all the materials were handmade rather than established measures. That project did not mean anything to me, and I completed it to get a grade. It might amuse though not surprise the reader that it did not help me get an A. When I challenged the professor that the grade was too harsh, he replied, "Well, next time, I hope you try harder." My pleas justly fell on deaf ears.

The following year, I completed an advanced research course for students intending to go to graduate school in which the professor challenged me to complete a publishable study. For my project, I took surveys to five churches of different denominations using the Basic Value Survey (Schwartz, 1992). I was testing a meaningful hypothesis that denominations would yield people with distinct patterns of the 12 values measured. Of course, I was naïve to make a causal hypothesis with an observational study, but the question still has value today.

The professor suggested that we work hard to write up the paper for publication, since I got a good sample size of "real" people, and our findings were consistent with predictions. We presented the findings at my first-ever professional conference, South Eastern Psychological Association (SEPA) (Grahe & Nelson, 1992), and I kept that paper to remind me of my final work as a college student.

It might further amuse readers that this was one of the first papers I completed using a computer, and I had to use the library computers, because only very rich people owned their own. Two nights before the final paper was due, my floppy disk was corrupted with a computer virus, and my file would not open. I had no backup except for the last draft I had worked on from about a week earlier. My professor offered me two lessons when I informed her of the situation. First, back up computer files. Second, it is easier to do something the second time.

As is the case with so many of my own students, I focused on a summer job and then moved to another state to attend graduate school. My professor had other things to focus on, and I could not do it alone. The data were not kept permanently, and so the value of that project was only for my own professional development; it did not benefit my professor; and without it getting into a journal, it failed to benefit the field either. Of course, this project did benefit me greatly compared to the first one. I read more theory while asking a question that was meaningful to me. I computed results using previously validated constructs using a reasonable sample size. Finally, the professor's dedication to taking the class to SEPA so that we could present our research in a professional setting introduced us to the field and showed us what our futures could include.

Years later when I started teaching research methods as an assistant professor, I framed projects in my own course more like the advanced research class I took rather than the methods course. It was this preference for authentic research experiences (Grahe, 2018) in which the project is publishable that eventually brought me to try to change research methods courses for everyone.

What's Wrong With Traditional Research Methods Courses?

I have two complaints about the structure of traditional methods courses (bad projects yield less learning and these courses are too cost intensive to waste). Primarily, there is overwhelming evidence that research experiences benefit students, and traditional methods courses do not yield projects that match the learning goals. Research methods courses that ask only for proposals and do not ask for data collection and analyses are not offering students the full array of research experiences. Projects in which students generate their own hypotheses rarely meet the threshold of being able to draw a strong inference (Grahe & Hauhart, 2013; Wagge et al., 2019). Small sample sizes, bad materials, poor delivery, and insufficient expertise are only a few impediments to students creating good research. Students end up with data that yield no firm conclusions, and so students are challenged to explain null findings in a theoretical framework when it was methodological limits that caused the lack of outcomes.

My second complaint is that these courses are too cost intensive to waste the experience. In a given year, methods instructors might mentor 5, 10, 15, 20 projects as part of their teaching load. At the same time, they are expected to publish research to achieve tenure and promotion. At medium and small institutions, these two competing research efforts can cause scarcity of research participants, let alone limited research capacity for the instructor. To the degree that these competing resources can be merged, institutional resources in addition to instructor and student resources can be aligned, and the problem of scarcity is instead replaced with increased benefits.

Updating the Curriculum With CUREs

Notice that I did not complain about the curriculum or content of the courses. Instead, my goal to improve the methods course has always fixated on the methods project. It is good to know that I was not alone in my belief that a more meaningful research project would lead to better learning. In the field of biology, Auchincloss et al. (2014) defined and studied outcomes related to course-based undergraduate research experiences (CUREs). CUREs are based within a course rather than an external learning exercise. Much like my own emphasis on developing research projects within the existing courses that were publishable, a core characteristic of CUREs is the same. When comparing CUREs to other laboratory/research experiences (i.e., traditional, capstone, and internship), CUREs are more diverse in structure. For instance, research questions and study materials might come from students or the instructor. Across five domains of the research process (practicing science, discovery, broader relevance, collaboration, iteration), CUREs not only deeply engage the student in learning, but they offer long-term benefits of participating in the publishing process. While natural sciences such as chemistry and biology can conduct research with sufficient power to address their questions in a single academic term, it is not the case in psychology and other social sciences. The need for large sample sizes and the existence of diverse populations makes it difficult for me to trust any single sample as being sufficient to draw strong inference for a research question. Instead, I prefer a large collection of samples.

CUREs: Great for One, Better With a Crowd

Early in my career, while teaching a statistics course, I read a call for instructors to join Alan Reifman in studying school spirit across many institutions.

In responding to his call, students in 22 classes collected a set of institutional (amount of alumni giving) and individual (people displaying the school logo) measures. In addition to knowing that we were contributing to data intended for publication (School Spirit Study Group, 2004), students learned about different units of analysis and nested designs in a very practical manner. When he sent another call inviting authors to help him study emerging adulthood and political attitude for the 2004 election, I again responded to the call. However, he also invited anyone who wanted to participate to add measures to his survey. The survey ballooned from two measures plus demographic items to a nine-page pencil-and-paper survey. We did not have computer software to collect data way back in the early turn of the millennium. Anyway, from the more than 20 instructors who originally committed to collect samples, only 10 actually submitted samples. Coupling this decreased participation with a failure to find support for the primary hypothesis was a minor disappointment at the time. However, that project offered an unusual opportunity. With an invitation to the readers of the *Emerging Adulthood* journal, a special issue including nine manuscripts reporting distinct findings developed (Reifman & Grahe, 2016).

The other outcome of participating in those projects was that I learned the value of undergraduate researchers. More importantly, I saw what a small group of faculty and students could do. When I experienced my own crisis of confidence in the field of psychology, it was undergraduate students who offered the solution. In Grahe et al. (2012), we argued that students could help test hypotheses by collectively focusing on research questions in their classes. While others built crowd projects to explore the replication crisis primarily at labs directed by Ph.D. researchers, I focused on projects for students. As the Reproducibility Project: Psychology (Open Science Collaboration, 2015) tested 99 hypotheses from prior research, students started testing hypotheses in their classrooms via the Collaborative Replications and Education Project (CREP; http://osf.io/wfc6u; Grahe, Brandt, IJzerman, & Cohoon, 2014). To augment the CREPs experimental and replication focus, I also built a survey project called the Emerging Adulthood Measured at Multiple Institutions 2: The Next Generation (EAMMi2; http://osf.io/te54b/; Grahe et al., 2018).

While the professional projects garnered media attention (Yong, 2019; https://osf.io/ezcuj/) and developed into a series of high-powered meta-science projects (Many Labs 1, Klein et al., 2014; Many Labs 2, Klein et al., 2018; Many Labs 3, Ebersole et al., 2016; Many Labs 4, Klein et al., 2019; Many Labs 5, Ebersole et al., 2020), hundreds of students contributed to a growing set of samples that contributed to a series of publications in recent years for both the CREP (Leighton, Legate, LePine, Anderson, & Grahe, 2018; Wagge et al., 2019) and the EAMMi2 (Faas et al., 2020; Grahe et al., 2018; Chalk, Barlett, & Barlett, 2020). What is particularly nice is that students could still participate in the professional projects if they were part of a lab that was participating. My students participated in four Many Labs projects (2, 3, 4, and 5). What amuses me to no end is that in 2010, there were zero options for instructors to choose from. But a decade later, I cannot sufficiently describe the breadth of options available in the space available. The next chapter explains these a bit more, and how to find other projects.

Undoubtedly, the way students experience their undergraduate research training has changed. Motivated students can gain so much more from their projects than I could have. Not only do these scaffolded research projects offer complex theory and hypotheses to decipher, but there are real opportunities for presenting research at a conference or even publication. For students who do not see those opportunities as valuable, they can still participate in meaningful research and take away the benefits known to emerge from undergraduate research experiences.

CHAPTER 7: CRISIS SCHMEISIS JOURNEY BOOK REVIEW

The Magnificent Mountain Women

Janet Robertson

The author of this book, Janet Robertson, wrote this partly because of her own passion as a mountain climber. She wanted to uplift the stories of women in the Rockies since most stories fail to recognize their presence. The book includes detailed stories about women with different reasons to go be in the Rocky Mountains. For those unfamiliar with this range, there are many 14,000-foot peaks that are desired destinations for climbers. The winters are brutally cold, and the summers are sunny and pleasant. Still today, it is a foreboding geography; 150 years ago, it was even more isolated and challenging. This is one of the books that I read specifically focused on women. I enjoyed this one very much because of the diversity of women represented in the pages and because the stories helped me imagine myself in their places. Besides the diverse perspectives, this book illuminates the challenges these women faced from discrimination and stubborn societal expectations. And their connections to the land and the stories of the mountains highlight the issues associated with environmentalism and sustainability (https://osf.io/tz6e7/).

Chapter 7 Exercises

Exercise 7.1: Connecting Experiences to Resumes: Identifying Job Skills From Coursework

- Make a list of skills collected as part of psychology major (writing, project management, critical thinking, etc.).
- Read this short article by Paul Hettich (2016) about how to recognize skill development while earning a psychology degree:
  - www.psichi.org/page/211EyeFall16cHettich#.YCGz2WhKi70
- Compare your list to the list of skills presented in this article.

- Where there are discrepancies, is it something you missed when you made the list or something that you have missed so far in your education?
- Make a priority list of skills that you intend to further develop for careers.

Exercise 7.2: Reasons to Embrace Open Science

The Second Stringers song suggests three reasons that people embraced the movement. However, there are many diverse reasons to engage in scientific transparency. Identify the three mentioned in the song in addition to identifying other reasons as well.

Exercise 7.2: Personal Research Philosophy

Using the ethical guides cited in this chapter, write a personal research philosophy. Essay prompt: As a researcher, I believe _____,

END-OF-CHAPTER REVIEW QUESTIONS

1. Describe the "publish or perish" environment created by many universities.
2. List three alternative Ph.D.s that a psychology undergraduate could earn.
3. List three skills a psychology degree can give you for a career in general.
4. What percentage of psychology majors go on to earn a Ph.D.?
5. How does completing a research project benefit undergraduate students?
6. What are ways in which undergraduate research experiences can be improved?
7. Name a few nonpsychology careers in which having a psychology degree could be a great benefit. Why?
8. How do undergraduate research opportunities help achieve goals in each of the facets of diversity, justice, and sustainability?

References

Auchincloss, L. C., Laursen, S. L., Branchaw, J. L., Eagan, K., Graham, M., Hanauer, D. I., . . . Dolan, E. L. (2014). *Assessment of course-based undergraduate research experiences: A meeting report*. Retrieved from https://www.lifescied.org/doi/full/10.1187/cbe.14-01-0004

Bangera, G., & Brownell, S. E. (2014). Course-based undergraduate research experiences can make scientific research more inclusive. *CBE—Life Sciences Education, 13*(4), 602–606.

Bauer, K. W., & Bennett, J. S. (2003). Alumni perceptions used to assess undergraduate research experience. *The Journal of Higher Education, 74*(2), 210–230.

Chalk, H. M., Barlett, C. P., & Barlett, N. D. (2020). Disability self-identification and well-being in emerging adults. *Emerging Adulthood, 8*(4), 306–316.

Ebersole, C. R., Atherton, O. E., Belanger, A. L., Skulborstad, H. M., Allen, J. M., Banks, J. B., . . . Nosek, B. A. (2016). Many labs 3: Evaluating participant pool quality across the academic semester via replication. *Journal of Experimental Social Psychology, 67*, 68–82.

Ebersole, C. R., Mathur, M. B., Baranski, E., Bart-Plange, D. J., Buttrick, N. R., Chartier, C. R., . . . Szecsi, P. (2020). Many labs 5: Testing pre-data-collection peer review as an intervention to increase replicability. *Advances in Methods and Practices in Psychological Science, 3*(3), 309–331.

Faas, C., McFall, J., Peer, J. W., Schmolesky, M. T., Chalk, H. M., Hermann, A., . . . Grahe, J. (2020). Emerging adulthood MoA/IDEA-8 scale characteristics from multiple institutions. *Emerging Adulthood, 8*(4), 259–269.

Fleischmann, M., Conroy, J., Christidis, P., & Lin, L. (2019, December). Datapoint: Psychology degrees build useful skills. *APA Monitor, 50*(11). Retrieved January 28, 2021, from https://www.apa.org/monitor/2019/12/datapoint-skills

Grahe, J. (2018). *Another step towards scientific transparency: Requiring research materials for publication*. Retrieved from https://www.tandfonline.com/doi/full/10.1080/00224545.2018.1416272

Grahe, J. E., Brandt, M., IJzerman, H., & Cohoon, J. (2014). Replication education. *APS Observer, 27*(3).

Grahe, J. E., Chalk, H. M., Alvarez, L. D. C., Faas, C. S., Hermann, A. D., & McFall, J. P. (2018). Emerging adulthood measured at multiple institutions 2: The data. *Journal of Open Psychology Data, 6*(1).

Grahe, J. E., & Hauhart, R. C. (2013). Describing typical capstone course experiences from a national random sample. *Teaching of Psychology, 40*(4), 281–287.

Grahe, J. E., & Nelson, L. J. (1992, April). *Value priorities of members of religious denominations: Catholic, Lutheran, United Church of Christ, and assembly of God*. Presented at the South-Eastern Psychological Association, Knoxville, TN.

Grahe, J. E., Reifman, A., Hermann, A. D., Walker, M., Oleson, K. C., Nario-Redmond, M., & Wiebe, R. P. (2012). Harnessing the undiscovered resource of student research projects. *Perspectives on Psychological Science, 7*(6), 605–607.

Hernandez, P. R., Woodcock, A., Estrada, M., & Schultz, P. W. (2018). Undergraduate research experiences broaden diversity in the scientific workforce. *BioScience, 68*(3), 204–211.

Hettich, P. (2016). Program your GPS: Guidelines to proficiency in skills for work and career. *Eye on Psi Chi, 21*(1), 20–24.

Karukstis, K. K. (2010). Multiple approaches to transformative research. In K. K. Karukstis & N. H. Hensel (Eds.), *Transformative research at predominantly undergraduate institutions* (pp. 21–34). Washington, DC: Council on Undergraduate Research.

Klein, R. A., Cook, C. L., Ebersole, C. R., Vitiello, C., Nosek, B. A., Chartier, C. R., . . . Ratliff, K. (2019). *Many labs 4: Failure to replicate mortality salience effect with and without original author involvement*. Retrieved from https://psyarxiv.com/vef2c/

Klein, R. A., Ratliff, K., Vianello, M., Adams, R., Bahník, S., Bernstein, M., . . . Nosek, B. (2014). Data from investigating variation in replicability: A "many labs" replication project. *Journal of Open Psychology Data, 2*(1).

Klein, R. A., Vianello, M., Hasselman, F., Adams, B. G., Adams, R. B., Alper, S., . . . Sowden, W. (2018). Many labs 2: Investigating variation in replicability across samples and settings. *Advances in Methods and Practices in Psychological Science, 1*(4), 443–490.

Landrum, R. E. (2018). What can you do with a bachelor's degree in psychology? Like the title, the actual answer is complicated. *Psychology Student Network*. Retrieved from www.apa.org/ed/precollege/psn/2018/01/bachelors-degree

Landrum, R. E., & Harrold, R. (2003). What employers want from psychology graduates. *Teaching of Psychology, 30*(2), 131–133. Retrieved from https://static1.squarespace.com/static/5681703a9cadb6554dbf0c78/t/56f6a32bb2d7c7a56b4aae74/1459004211758/What+employers+want+from+psychology+graduates+%28Landrum+%26+Harrold%2C+2003%29.pdf

Leighton, D. C., Legate, N., LePine, S., Anderson, S. F., & Grahe, J. (2018). Self-esteem, self-disclosure, self-expression, and connection on Facebook: A collaborative replication meta-analysis. *ResearchGate*. doi: 10.24839/2325-7342.JN23.2.98

Lopatto, D. (2004). Survey of undergraduate research experiences (SURE): First findings. *Cell Biology Education, 3*(4), 270–277.

Nosek, B. A., Spies, J. R., & Motyl, M. (2012). Scientific utopia: II. Restructuring incentives and practices to promote truth over publishability. *Perspectives on Psychological Science, 7*(6), 615–631.

Open Science Collaboration. (2015). Estimating the reproducibility of psychological science. *Science, 349*(6251).

Reifman, A., & Grahe, J. E. (2016). Introduction to the special issue of emerging adulthood. *Emerging Adulthood*, 135–141.

Russell, S. H., Hancock, M. P., & McCullough, J. (2007). Benefits of undergraduate research experiences. *Science, 316*(5824), 548–549. doi:10.1126/science.1140384

School Spirit Study Group. (2004). Measuring school spirit: A national teaching exercise. *Teaching of Psychology, 31*(1), 18–21.

Schwartz, S. H. (1992). Universals in the content and structure of values: Theoretical advances and empirical tests in 20 countries. In *Advances in experimental social psychology* (Vol. 25, pp. 1–65). New York: Academic Press.

Seymour, E., Hunter, A. B., Laursen, S. L., & DeAntoni, T. (2004). Establishing the benefits of research experiences for undergraduates in the sciences: First findings from a three-year study. *Science Education, 88*(4), 493–534.

Taraban, R., & Logue, E. (2012). Academic factors that affect undergraduate research experiences. *Journal of Educational Psychology, 104*(2), 499.

Wagge, J. R., Brandt, M. J., Lazarevic, L. B., Legate, N., Christopherson, C., Wiggins, B., & Grahe, J. E. (2019). Publishing research with undergraduate students via replication work: The collaborative replications and education project. *Frontiers in Psychology, 10*, 247.

Yong, E. (2019, November 12). *Psychology's credibility crisis*. Retrieved February 9, 2021, from www.discovermagazine.com/mind/psychologys-credibility-crisis

8 OPEN SCIENCE ALPHABET

Learning to Read

Chapter 8 Objectives

- Describe why researchers should avoid initials in writing
- List and describe the projects sponsored by organizations
- Explore projects that invite a crowd
- Show ways that a student can serve as a collaborator
- Discuss other open-access resources to streamline work

MUSIC IN A BOX: ABOUT THE SONG "THE OPEN SCIENCE ALPHABET"

"The Open Science Alphabet" is a false closer with a big ending that rouses the listener with invitations to sing along. It is intended to review the many ways researchers can participate in collaborative, replicative, and transparent science while encouraging the listener to keep active in learning about new open science initiatives. A second message, which is important for the members of the movement to remember, is that using acronyms and initials can lead to feelings of ostracism and exclusion. It also violates APA writing style.

OPEN SCIENCE ALPHABET: LEARNING TO READ

One of my earliest memories is going to a drive-in movie and seeing *The Sound of Music*. I wonder if that is the movie I saw and I wonder if the adult who took me was my aunt. It is a reminder that we should not trust our memories that I share that memory without confidence. I also share this memory because one of the lead characters, a nun named Maria, introduces the children in her charge to singing with a song, "Do Re Mi." In this song, she explains that while letters make up the building blocks of language, the building blocks of song are contained in notes. The premise of the song is that once a singer understands the steps of a scale and knows how to reference them with words (do, re, mi, etc.), they can build any song. After all these years, this movie still moves me emotionally when I see it, and that song reminds me of the joys of music. As the scope of open science becomes more complex, researchers and consumers of research need to navigate the scattered lexicon of opportunities. Much like singers open up the world of music if they understand the relationship between steps in a musical scale and are able to use language to communicate about those steps, researchers open up the world of science if they understand the basic tenets of scientific transparency and know how they relate.

Throughout the book to this point, prominent projects and initiatives were introduced, sometimes briefly. This chapter will serve as a final organization and summary of those projects and initiatives. It will also offer some considerations when communicating about open science initiatives with others. These considerations can be seen through other connections between the alphabet metaphor and open science, which are relevant here. One connection is amusing and inviting, while the other is threatening and exclusionary. Further, addressing these promptly before trying to organize and summarize will assist in the process.

The amusing connection that highlights an inviting component to open science language is the way the Center for Open Science structured the Open Science Framework webpage names. When they first began, the stem of any project weblink was http://openscienceframework.io/, and each individual project was randomly assigned five characters (e.g., Reproducibility Project: Psychology; RP:P, https://osf.io/ezcuj/). The "ezcuj" section of the weblink represents a random letter or number combination that is associated with the particular project. As evident by the present address for the RP:P, they shortened the stem to http://osf.io/. For me, this cemented the most inviting open science website nomenclature possible. Given that short-term memory capacity is seven +/- two chunks of information, even five randomized digits are easily remembered. Thus, in addition to OSF.io being both short and easy to remember, the basic project page is similarly easy to remember. This helps to invite others to projects, because even if there was no way to write down the basic information, an invitee only needs to remember five digits to get to a project later.

The threatening and exclusionary connection may surprise some readers because it is such an integral part of common communication strategies: our predilection for using initials to represent names and phrases. This practice is problematic because it is ethnocentric, assuming that everyone has the same historical references. Further, it causes feelings of ostracism (Hales, Williams, & Rector, 2017) which leads to feelings of disconnection, lower self-esteem, or even anger. It is also very difficult to process manuscripts that use acronyms and initials (Gernsbacher, 2013). In contrast to the short website address that encourages short-term memory encoding, employing initials and extensive jargon discourages new contributors and signals an indifference to individuals beyond the existing in-group. APA writing guidelines (look at me referencing APA and not American Psychological Association) stress that writers should avoid any initials unless they are commonly known and understood. Should a writer need to introduce initials, they are obliged to first present the full name. Thus APA allows for initials in some cases, but Gernsbacher (2013) points out that the length of manuscripts demands reminding the reader of initials more frequently than once and that the benefits from reduced keystrokes and printed pages are far outweighed by the costs of alienating a reader by continue to use initials.

NOT ALL POPULATIONS ARE KNOWN: FINDING OPEN SCIENCE RESOURCES

When I first started recruiting students and faculty into the world of open science, I tried to offer a complete list of all current and past open science initiatives. I invited the audience to let me know if anything was missing. It did not take long until the pace of new opportunities quickened beyond my capacity to share the information in a single slide or a few minutes of a talk. The momentum of the open science movement has not slowed, so the idea of collecting a complete compendium of all the important projects, tools, and resources is more elusive now than ever. Unlike a goal such as visiting all 50 states or all national parks in the United States, where a list is either static or officially updated with new entries, there is no central location for all open science. Thus, this chapter will offer an estimate of the population of open science initiatives rather than a complete list. It is organized into sections representing (a) organizations, (b) projects, and (c) resources. As with all samples, this one contains bias. The examples are primarily for psychology, they tend toward those with greatest advantage for students and faculty, and they are those with which I am most familiar. However, with familiarity with these, researchers should be able to find others as they arrive in the future or in other areas.

Organizations Supporting Transparency

The first formal organization dedicated to open science, the *Berkeley Institute for Transparency in Social Sciences (BITSS)*, was founded in 2012. BITSS invited researchers and educators across disciplines to embrace open science through workshops and targeted grants. Researchers committed to advancing open science can apply to become BITSS catalysts, external voluntary affiliates who conduct programming and contribute to projects. Beyond their yearly conference at their host institution at the University of California: Berkeley, the organizers and BITSS catalysts offer multiple open science trainings called Research Transparency and Reproducibility Training (RT2) workshops. To date, these grants have funded 26 Social Science Meta-Analysis and Research Transparency projects that all (a) develop new methods to improve science, (b) study methods or develop tools to improve meta-analysis techniques, or (c) study the culture of science as it responds to open science. One resource of particular value is the compendium of open science resource library (www.bitss.org/resource-library/). Topics in 17 categories represent all facets of the research process (planning, 7 categories; collecting and analyzing data, 4 categories; dissemination, 6 categories). As their names suggest, BITSS represents many academic disciplines including those in the social sciences, but they also have representation in other areas such as humanities and engineering.

The Center for Open Science (COS; http://cos.io/) emerged shortly after BITSS with many goals, but their primary impact was to disseminate new software to help manage the research workflow called the Open Science Framework (OSF; http://osf.io). Beyond developing and maintaining the OSF, they have resources aimed at authors, publishers, and funders of research. In addition to written instructions and guidelines for things such as Open Science Badges (http://osf.io/y2hjc), Preregistration (https://osf.io/prereg/), and Transparency and Openness Promotion guidelines (https://osf.io/9f6gx/), they have their own YouTube channel (www.youtube.com/channel/UCG-PlVf8FsQ23BehDLFrQa-g) with hundreds of hours of tutorials and helpful advice. The director and co-founder of the COS, Brian Nosek, is perhaps the most easily recognized figure in open science. He helped organize the Reproducibility Projects and served as a primary researcher for all the Many Labs projects. The COS welcomes all motivated researchers of any level, and there is often volunteer work for new projects. Individuals who become proficient on the OSF can become a COS ambassador to help train others to use their free online management software, which is dedicated to encouraging rather than threatening researchers into complying with openness.

The Center for Open Science was organized by a research psychologist but intends to reach all disciplines. For instance, their OSF software is usable for any scholarship, and the TOP guidelines speak to transparency across disciplines. Additionally, they also develop projects such as the RP: Cancer (www.cos.io/rpcb), which tests replication in medical sciences. Finally, the Preregistration Challenge offered $1,000 prizes to any researcher who published a paper that included preregistration.

Both BITSS and the COS are organizations located in physical locations that derive funding from grants and donations. Their missions are multidisciplinary and their organizations focused on collecting and disseminating resources. In contrast, the Society for the Improvement of Psychological Science (SIPS) is a new organization that is driven by membership rather than centralized leadership. Though SIPS was founded by Simine Vazire, another easily recognizable open science advocate, leadership follows the elected members of the executive board. Before speaking more about the organization, Simine Vazire deserves a bit of extra recognition as many of her opinion pieces and commentaries early in the movement were easy to digest and offered strong arguments for the need to bring about change. She heroically challenged norms in existing societies such as the Association for Psychological Sciences and the Society for Personality and Social Psychology while editing their prestigious journals and demanding better science from the submitting authors. She eventually founded SIPS to collect all those passionate on the topic within the field.

In SIPS, rather than try to develop ideal publishing expectations in existing journals, they demonstrate their ideal with their own journal: *Collabra* (discussed later in this chapter). The first few SIPS conferences had such high demand that attendance was closed only a couple days after registration opened. These small conferences included sessions called "hackathons" in which attendees would brainstorm common solutions. These sessions extend beyond the conference, and attendees craft a manuscript for submission to disseminate their assessment and solution of transparency problems. Thus far, they have published on topics such as effects of the COVID-19 pandemic (Rissman & Jacobs, 2020; Sætrevik, 2021), accessing open science (Kathawalla, Silverstein, & Syed, 2021), and language usage (Kleinberg, Nahari, Arntz, & Verschuere, 2017; Sassenhagen et al., 2018).

RESEARCH IN ACTION: THE NATIONAL PARK SERVICE AND OPEN SCIENCE

With locations all over the United States, the National Park Service is a living laboratory that samples some of the unique geographical and geological features in the country. For the professional researcher, there are numerous projects occurring at any time. Many of these have been published in the Open Science DB (www.opensciencedb.com/), which hosts the projects of many government agencies. "The National Park Service (NPS) values and uses citizen science in pursuit of its mission to preserve natural and cultural resources and to provide enjoyment and education for visitors" (National Park Service, 2020). Through the National Park Service website, one can browse their current citizen science topics (environmental change, biodiversity, culture and heritage, events) to find a project of interest. While national parks may seem more suited to physical science, there is plenty for social scientists as well, from the history of how people used the land to how people interact with and use the parks today. More than half of the jobs and internships listed on the website include research that will eventually be posted on an open science platform. Projects range in timeframe from a few hours of one day to many weeks. This is the perfect place for an undergraduate student to get experience at the parks and help with crowdsourced, open science without the obligation (or pressure) to be the one publishing the findings.

OPEN SCIENCE PROJECTS INVITING A CROWD

One of the most notable changes in psychology research over the second decade of the 21st century was the emergence of crowdsourcing as a tool to answer research questions and conduct meta-science. As the decade came to a close, Uhlmann et al. (2019) published a paper outlining the various benefits of using crowds of researchers to advance science. To conceptualize the comparison between traditional and crowdsourced models, consider the comparison between vertical and horizontal organizational models. A vertical approach accomplishes a task with the same organizational structure throughout the process. In the case of the traditional research model, that structure included a small group of researchers, often with a single leader. They note its characteristics as being localized and resource intensive, generating many instances of small science with authorship as the primary reward. They assert that the horizontal approach differs as a continuum rather than being qualitatively different.

Compared to the ultimate vertical model with a solo researcher working in isolation, crowd projects have decisions, costs, and rewards distributed across the contributors. Rather than many instances of small science, distributed costs and labor enable crowds to conduct big science. Questions that were not possible before are now routinely asked using large crowds of researchers. With many different types of crowd projects, the rewards are also more diverse than manuscript authorship and grant funding. Further, Uhlmann et al. review ways that crowds can be leveraged across the entire research process (ideation, assembling resources, study design, data collection, data analysis, replicating research to publication, writing research reports, peer review, replicating published findings, deciding future directions). Because the list of crowd projects is too extensive to be reviewed exhaustively, I summarize the history and status of a few projects that are most relevant to students of psychology. Most of these have been referenced in some fashion already in this book and are presented here to allow readers to understand projects with historical relevance as well as those with ongoing critical value. Finally, this summary will highlight some critical differences in project structures.

Comparing the Early Entries

In January 2012, before the Replication Crisis was so named, two crowd projects were announced in a special section in the *Perspective on Psychological Science* journal (the Filedrawer Project and the Reproducibility Project: Psychology). They offered distinct and novel approaches to addressing replications in research. The Filedrawer project (Spellman, 2012; www.psychfiledrawer.org/TheFiledrawerProblem.php) took its name from a term Rosenthal (1979) used to describe the phenomenon introduced earlier in this book whereby researchers put their failed replications and other studies into a "file drawer" because they either could not or did not want to get them published. This project established an online repository for those studies with the hopes that the data sets would be useful in meta-analyses over time. This project represents a mostly decentralized organization with no oversight regarding what topics are studied or the methods employed (direct replications, close replications, novel research). In contrast, the Reproducibility Project: Psychology (Open Science Collaboration, 2012, 2015) represented a more centralized approach. Rather than inviting any data, the Open Science Collaboration was a large crowd of 270 researchers that attempted to replicate 100 effects published in three prominent journals (*Journal of Personality and Social Psychology, Psychological Science, JEP: Learning, Memory, and Cognition*). Though the crowd was dispersed across the world eventually reaching 270 contributors, a central authority (eventually located in the Center for Open Science) worked to ensure methodological compliance.

The RP:P was a lightning rod of excitement from both admirers and critics, particularly after the published results supported the original conclusions of less than half of the replications (Open Science Collaboration, 2015). Van Bavel, Mende-Siedlecki, Brady, and Reinero (2016) argued that the findings were related to the context under which the replication was conducted. They identified more successful replications to be more closely matching the original findings. Other concerns identified small sample sizes (Etz & Vandekerckhove, 2016) or distinct populations (Gilbert, King, Pettigrew, & Wilson, 2016) as the reasons for inconsistencies with original findings. Finally, there were criticisms that expertise of the researcher predicted outcomes (Bench, Rivera, Schlegel, Hicks, & Lench, 2017; Klein et al., 2019; Ebersole et al., 2020a). While there were many criticisms of the conclusions, the field of psychology could not ignore the implications.

Many Labs Do It Better Than One

I was personally amused by some of the criticisms because the project was so obviously a pilot test. In fact, while RP:P contributors developed their materials and collected their data, Brian Nosek was overseeing Rick Klein's development of the first Many Labs project (Klein et al., 2014). The major methodological innovations of Many Labs compared to the RP:P was the increased centralization of both study selection and materials and methods uniformity. Rather than helping contributors develop replication materials, the Many Labs team shared one set of centrally developed materials. One outcome of this change was a swifter conclusion of the project, which published its findings a full year before the RP:P even though it started more than a year later. Notably, most of the replications yielded statistically significant outcomes (10 of 13) even though the release of the findings created conflict and controversy. In a flurry of angry social media posts, some started using demeaning terms (e.g., second stringers, data pirates, replication bullies, etc.) and accusing them of trying to attack peers by undermining their careers through targeting certain effects. Replicators should accept responsibility for this, as some of their own social media posts regarding the replicability were sometimes coarse and inconsiderate. It was in this mindset that I argued in my own social media post that we should avoid engaging in a "witch hunt" as we worked through the scientific transparency revolution (Grahe, 2014). Naturally, science advanced through evidence from data, not opinions on social media.

It is worth noting that Brian Nosek continued driving the Many Labs projects to address criticisms of the first Many Labs outing and the Reproducibility project. In Many Labs 2 (Klein et al., 2018), the slate of studies was expanded from 13 to 28, and the number of labs expanded from 36 to 125. In Many Labs 3, Ebersole et al. (2016) narrowed the question of replication to whether effects were impacted by time of the semester they were tested and used 20 labs to study 10 effects. Many Labs 4 (Klein et al., 2019) tested a single effect (terror management theory, Greenberg, Pyszczynski, Solomon, Simon, & Breus, 1994) to determine if experimenter expertise impacted outcomes. Across locations, participants completed the study either using materials developed by the experts, including original authors, or developed by the researchers in the crowd. Many Labs 5 (Ebersole et al., 2020b) similarly examined expertise but targeted reviewer expertise rather than experimenter expertise. Addressing concerns raised about RP:P replications, Many Labs 5 targeted 10 effects that were criticized as being poorly administered or underpowered using multiple locations (ranging from 3–9 labs per original effect; median = 6.5).

The summation of the Many Labs findings might be considered a bit disappointing, as they regularly found estimates that were smaller or non-existent than prior studies, even when selecting studies explicitly with the expectation that they would replicate (see Many Labs 3 and 4). Stroebe (2019) argues that we learn relatively little from these studies because they do not help reveal fraud, their lackluster findings suggest the need to identify contexts of effects rather than reveal false findings, and they do not test or advance theory. In contrast, these should be seen as early attempts to effectively test questions using large-scale crowd projects. Each Many Labs advanced our knowledge by offering new estimates of established effects, and each advanced methodologically compared to the previous going back to the RP:P. I estimate that these Many Labs projects might occur again, but for now, they are dormant. The ultimate success of these projects might have been to collect motivated and like-minded researchers so that they could create the next generation of crowd projects that I discuss at the end of the section, the Psychological Science Accelerator (Chartier, 2017; Moshontz et al., 2018).

Though the Many Babies project sounds like it is related and shares much of the approach of the Many Labs franchise, the project has a different founder, Michael Franks, and is not directly connected to the COS, and the target population is infants rather than adults. Finally, Many Babies studies are not purely for replication purposes but rather to advance theory. If other disciplines struggle to obtain large sample sizes, the challenge is virtually insurmountable when studying infants and children. The extra layers of ethical concerns for participants under 18 is compounded by the lack of some centralized space to find infants and toddlers. As such, the benefit of crowd projects offering researchers more opportunity to conduct meaningful research in resource-poor locations is amplified by many factors when considering infant research. In the 4 years since its inception, Many Babies has started five projects, with one already resulting in a publication (Byers-Heinlein et al., 2020), six spin-off projects, and two secondary analyses (https://manybabies.github.io/projects/). I predict that Many Children and Many Adolescents projects are merely waiting for a champion to develop them as peers watch Many Babies continue to excel and reward its contributors.

There are other types of "many labs" projects that focus on answering a single question and then disperse. Pipeline projects (Schweinsberg et al., 2016) and Registered Replication Reports (Simons, Holcombe, & Spellman, 2014) represent end points on harnessing large numbers of labs to replicate some effect. For Pipeline projects, the goal is to test the methods supporting a theoretical assertion before the first publication of that effect. Schweinsberg et al. chose 10 effects associated with moral judgments selected by the planning group.

Twenty-five labs replicated the effects to determine which effects were strong before publication. In this case, they found 6 of 10 to be reproducible.

This unusual approach to publication is still very rare, with only one published example and only one other active call (https://osf.io/skq2b/).

In contrast, Registered Replication Reports (RRR) focus on well-established effects. Defining characteristics of Registered Replication Reports are that editors are involved in the study planning, and many labs are used to achieve sufficient power to test the question. Unlike most other publication situations, RRRs are reviewed before the data are collected. The editors and authors must agree that a methodology will adequately test the hypothesis. If they agree on the methods, then they receive a conditional acceptance for publication, assuming that they meet any pretest and sample size criteria. If this sounds like a Registered Report, it is because the only defining characteristic that separates an RRR from an RR is the contribution from many labs. Where a Registered Report might occur with a single lab, any RRR requires many labs. Since the first RRR was published (Alogna et al., 2014), there have been many subsequent calls, partly driven by a number of journals dedicating pages to these types of studies (e.g., *Cortex*, *Collabra*, and *Perspectives on Psychological Science*). Though all of these projects offer contributors professional benefits associated with authorship, they mostly exclude student researchers unless that student is working as an assistant in a faculty lab. But there are projects dedicated to including students too.

Students as Crowd Contributors

The RP:P and Many Labs garnered international attention when they arrived, but the model of using a crowd to generate data was not new. Beyond psychology, examples come from the citizen scientist movements that have made advances in astronomy through signal detection (Science United, n.d.; SETI@ Home, n.d.), biology through migration (Citsci_admin, n.d.; Welcome to journey north, n.d.), and plant identification (Global Garlic Mustard Field Survey, n.d.). But even within psychology, the School Spirit Study Group (2004) used a crowd model in 2002, and the EAMMI did so again in 2004. Even in 2012, as the RP:P was announced, the Collective Undergraduate Research Project (Grahe, 2010) had already been soliciting contributors to join for more than a year. Why did they not receive the same level of excitement? Maybe because none of those projects were presented as a response to a crisis but rather as a mechanism to better use resources (Grahe et al., 2012). Additionally, these projects similarly included observational and survey rather than experimental methods, which limits the ability to draw causal inference. Alternatively, students also needed to learn to be researchers revealing some questions about students' capacity to successfully conduct research. Finally, adding authentic research opportunities to a course requires planning and extra work.

The Collaborative Replication and Education Project

CREP (http://osf.io/wfc6u/; Grahe, Brandt, IJzerman, & Cohoon, 2014) addressed some of these limitations by inviting students to replicate experiments published in top-tier journals from a curated list. The CREP selects studies by collecting the top-cited manuscripts in the top journals across nine psychology subdisciplines from 3 years prior (Wagge et al., 2019). Since its inception, hundreds of students completed the process to be CREP contributors. This included submitting projects for review before data collection and again after results are posted. Submissions include evidence of proper materials, ethics approval, and a video of the procedure to ensure high-quality data. When enough samples have accumulated to warrant a meta-analysis, the findings are submitted for publication. In this model, students make contributions to science while learning the craft of research. Depending on their motivation, students might present their single studies at a professional conference or serve as CREP administrators, who volunteer to help the CREP help other students. Truly motivated students participate in the authorship process.

Where the CREP offers students an ongoing, experimental approach to participate in the crowd of research, the EAMMi2 invited students to help in data collection for only one year but offers a data set for either study or scientific advancement. As a close and conceptual replication of the EAMMI, the EAMMi2 included measures of emerging adulthood, political attitudes, mindfulness, and demographics on a stem survey (Grahe et al., 2018). In addition, contributors could submit proposals for other measures to be added. A survey taking approximately 30 minutes and including 20 psychological measures was the result. Researchers (both faculty and students) at 32 locations invited peers and friends to complete the survey, with more than 4,300 starting the survey and 3,200 remaining in the cleaned data set. As with the original EAMMI, the EAMMi2 was part of a special issue for the *Emerging Adulthood* journal, and the data offers continued opportunities. Students can use the publicly available data set for practice, but they can also use it to test publishable novel or replication questions as secondary analyses.

The Network for International Collaborative Exchange

NICE is another student-friendly crowdsourcing project. Sponsored and maintained by Psi Chi, the International Honor Society for Psychology, the NICE invites students and faculty from across the world to conduct research with a cross-cultural focus. NICE projects could be experimental or survey in design, as they result from submitted proposals selected by the organizers each year. Starting in the summer of 2017, the NICE is relatively new, with data collection completed on one project and ongoing with two more.

Transformative Transparency: Psychological Science Accelerator

With an origin attributed to Chris Chartier during an aha moment while riding a bicycle (Chartier, 2017), the Psychological Science Accelerator (PSA) quickly demonstrated itself as the next step after Many Labs. The whirlwind development of the PSA is somewhat attributed to his participation in the RP:P and Many Labs projects. New members flocked to the PSA like ferrous metal drawn to a magnet. As interest grew into the hundreds, the PSA was able to decentralize into nine committees for study selection, data management, media relations, and other related functions. With stated goals of transparency, diversity, and inclusion (Moshontz et al., 2018), the PSA successfully recruited contributors from dozens of countries from all six continents where humans have permanent residences. Upper leadership positions were reserved for non-US contributors, and all decisions resulted from votes.

In the three brief years of their existence, the PSA has started six standard projects and three rapid research projects designed in response to the COVID-19 pandemic. Of these projects, three standard projects and all three rapid responses have completed data collection. The other three projects were interrupted by the pandemic but will likely restart by the time this book is completed.

I share the excitement of my peers that the PSA is the next stage of psychological research. The extensive focus on detailed procedures has created a system that should last beyond any single member. Rather than being driven by a single individual, the PSA moves by the will of its collective membership, a truly international and diverse collection of psychology researchers. In time, this project could transform the very nature of psychological research. For instance, after hundreds of contributors carefully selecting research questions, planning and executing studies, analyzing the data, and providing reports, how much value-added benefit is there from an editor and three reviewers in a journal submission? This question applies to the PSA more than the other crowd projects particularly because of the size and diversity of the crowd. Regardless of the answer to that question, research in psychological sciences will never be the same now that the value of harnessing the power of crowds has been realized.

Possible Uses of Past Projects

Understanding the historical significance, unique characteristics, and contributions of completed and ongoing crowd project types helps researchers find projects that are good matches for their own future efforts. Perhaps learning about these projects will inspire some new RRR or Many Labs project. In any case, there are many other possible uses for projects that completed data collection and reported their findings including (a) data reanalysis or secondary analyses, (b) population estimates for measures and effects, and (c) data training.

These projects and their data offer opportunities for secondary analyses or data reanalysis because there are many ways to consider results. As discussed in Chapter 5, there are many researcher degrees of freedom in data analyses. Many of those decisions do not have a single correct solution. Even something as seemingly simple as what to do with missing data has multiple solutions, as the researcher might complete pairwise deletion in which participant data is not included in variables where there is missing data or list-wise deletion in which the participant is removed from all further analyses. As the decisions and analyses get more complex, the potential impact on final conclusions becomes more varied. Which effect size estimate is the best to consider from any data set? There are better sets to consider, but not any single estimate could be considered the true estimate.

This variability in potential outcomes is called the multiverse of p-values (Steegen, Tuerlinckx, Gelman, & Vanpaemel, 2016). To complete a multiverse of p-value analyses, researchers conduct a set of analyses with varying data sets selected from following different decisions in coding or analysis (Silberzahn et al., 2018). By looking at the distribution of p-values, one can estimate how reproducible the effect is across some conditions associated with testing. Because this multiverse of results exists, any data set with sufficient value could be approached by contradictory researchers to get a more complete test of any hypotheses. For instance, there are examples of secondary analyses as criticisms of the RP:P, the various Many Labs projects, and some RRRs. The iterative nature of science requires debate about the conclusions we draw about data. Because crowd projects also tend to be open science in nature, the data tend to be available.

A second use would be to use them to help establish population estimates for common measures and effects. For instance, researchers should conduct a power analysis before collecting data to determine the ideal sample size for adequate power. It would be nice if there was a compendium of effect sizes somewhere that collected all previous work and summarized it quickly. While such a resource does not yet currently exist, these data sets are better estimates than any single published study because they are large and more diverse than most published research. Further, they tend to have smaller effects than published findings, even when the findings are significant. Thus, combining original published findings with estimates drawn from a crowd, the power analysis is more sensitive to the population and less impacted by publication bias, which tends to favor large effect sizes with small samples.

A third use could be combined with the first two, or not; these large complex data sets are perfect for training new researchers in how to conduct statistical analyses. The EAMMi2 data was explicitly intended to become a training data set in addition to a contribution to science. The project includes a broad array of measures with a large sample size that is nested across locations. Any statistical tool that is taught to undergraduates can be demonstrated in the data set. As such, the EAMMi2 is building a compendium of examples and exercises from various instructors that use it.

One of my favorite memories about this project was finding out that a student who graduated from our program was being taught graduate statistics from someone who found the EAMMi2 data set online. It is hard to track project impact through the use of teaching exercises, so it is very rewarding when a concrete example arises.

Another advantage of the EAMMi2 data, and the Many Labs 1, 2, and 3 data, is that their complexity allows for follow-up testing of novel questions. Instructors could invite students to use these data sets to learn how to conduct statistics and simultaneously learn how to conduct research. If the secondary analysis of a studied effect is interesting, the student could also learn how to navigate the publication process and author the paper. An example of what could be done can be seen in the Open Stats Lab (https://sites.trinity.edu/osl/instructors-using-osl-classroom) project, which offers basic statistics and methods tutorials using carefully selected published data sets. There are examples for correlation, regression, *t*-test, one-way, factorial ANOVA, and miscellaneous others. In total, the 17 examples are quite useful for introductory statistics. What the crowd projects could offer is the same level of learning for much more complex tools.

A POCKETFUL OF OPEN SCIENCE TOOLS

These examples of crowd projects and their possible uses represent only one type of open science initiative. Moreover, participating in crowd projects often requires capacity to work with tools developed to facilitate these projects. While I review a few personal favorites that are particularly advantageous for various parts of the workflow, the BITSS (www.bitss.org/) and COS (www.cos.io/) resource compendiums attempt more extensive coverage. Learning about these types of tools enables researchers to better critically evaluate others not reviewed here, particularly those that do not yet even exist. Here, I focus on tools that assist with communication, project administration, data management/analyses, and publication.

Free Learning With Open Access

Some confuse the concept of open access with open science. However, it is possible to share open science behind a paywall. For instance, journals that publish work with open data or open materials but still charge for access to the manuscript that publishes that data and materials is not open access. Rather, open science initiatives are of a broader interest in open access. While some argue that science is improved to the degree that all science is shared freely (Nosek, Spies, & Motyl, 2012; Nosek & Bar-Anan, 2012; Scientific Utopia 1 & 2), publishers still exist, and there are costs associated with accessing published work. And yet there are many open access resources in education so that a motivated researcher does not need deep pockets to access learning tools.

When conducting research, there are tools to help get to published research even without access to the journal. For instance, Research Gate (http://researchget.net) offers a service where researchers can post personal versions of their papers. For papers that authors do not have permission to share publicly, they can share personal copies via direct request in a messenger system. This and other preprint repositories provide greater access to both published and not-yet-published work. Preprint servers are available in many disciplines, but in psychology, the PsyArXiv preprint service offers an easy way to publish work (https://psyarxiv.com/). In some cases, it is a mechanism to get extra feedback from peers before submission. In other cases, it is a mechanism to share work that might be important but won't otherwise get published, such as graduate and undergraduate theses.

Communicating Online Freely With Google Tools

The suite of Google tools which accompany a basic Gmail account offers many advantages for research, particularly when working with crowd projects, but advantages resonate for any size research team. It is hard to remember the struggles of writing collaboratively before there was simultaneous online editing of a document. With varying access settings, researchers can share links that invite readers to view only or make comments or collaborators to edit. Combined with other tools, researchers can announce a new project by sharing a Google Doc link. At the bottom of the call, a link to a Google Form measuring interest and capacity collects critical contributor information, and the information is reviewed via a Google Sheet. With many plug-ins, Google offers a plethora of tools. And when the collection of researchers is assembled, Google Groups can be used to track communication and emails. While there are other similar services, Google is the cheapest to use.

Managing Projects Freely With the OSF Versus Other Repositories

What the many advantages in the suite of Google tools does not provide is a place to keep data and materials or other non-Google files. To manage this, a researcher needs a repository. Here, there are many options including dataverse (https://dataverse.org/) and figshare (https://figshare.com/), and many universities are creating their own to keep greater control of academic capital. However, the OSF is better than all of them. The OSF has been discussed multiple times in this book, so another review would be redundant. However, it is worth noting here that no other data repository offers the capacity to build anonymous view-only links. Though other repositories have flashier user interfaces, without view-only links these repositories are useful only at the end of the peer-review process. Further, other repositories primarily serve as a location, whereas the OSF can serve as another source of information because each component's description window and wiki can be used for extensive delivery of instructions and content. Entire crowd projects have been run through a combination of Google tools and the OSF (CREP, EAMMi2, Many Labs). In contrast, an OSF project is useful from the beginning to the end of the process, as hopefully demonstrated in this book.

Open Sesame and the Wave of Free Online Study Software

The Many Labs and EAMMi2 projects used Qualtrics programs to disseminate their experiments and surveys. While Qualtrics is very powerful, the user license requires extensive resources. Free alternatives are continuing to develop. Psycstudio (www.psychstudio.com/), LIONESS (https://lioness.uni-passau.de/bin/demo.ph), and Open Sesame (https://osdoc.cogsci.nl/)

all emerged in recent years as free online management tools. As this field is so young, it is worth reviewing all options before selecting one.

Analyzing Data Freely With Powerful Programs: JASP and R

Statistics software can be pricey. SAS, SPSS, and Matlab are incredibly powerful tools, but they require licenses with yearly fees in the hundreds of dollars. The R statistics platform (www.r-project.org) can do what any other program can do and more. Because it is an open-sourced and free software package, users are constantly updating and creating new statistics packages to augment basic R capabilities. The power of R is truly remarkable, and the fact that it is free is amazing. The challenge is that users must understand how to write R code in order to use R. The learning curve for R is steep, and thus people with R expertise can often find lucrative work as consultants if they are so motivated. Graphical user interfaces (GUIs) help the more common user by offering a point-and-click dialog box that then writes the R code. There are a number of GUI options for R, but open science advocates might be drawn toward JASP, since it has many default options that allow for testing Bayesian models. In any case, R is an open science tool because it allows anyone to conduct any statistics for free if they can master the program.

CONCLUSIONS

This chapter briefly summarized a large collection of open science tools including various crowd projects and other open science tools. After warning the reader that using too much jargon, initials, or abbreviations can lead to feelings of rejection, the chapter inundated the reader with an open science alphabet, so it was challenging to avoid violating exactly this advice. Open science advocates should be careful with their own communications about these projects and find more projects as new initiatives emerge. The open science alphabet just keeps getting more complete.

CHAPTER 8: CRISIS SCHMEISIS JOURNEY BOOK REVIEW

Indians in the Yellowstone Park

Joel Janeski

This book is written by Joel Janetski who is an anthropology professor and museum director who studies Native people of the American West. This book focuses on the various people who lived in the areas now known as Yellowstone National Park. I really enjoyed learning about the different groups who lived around this area and how they used it. The depressing nature of the interactions between Whites and the Indigenous peoples are recurring across this book. Never-the-less, this book connects to diversity, social justice, and sustainability and compels me to wonder about how we could do better moving forward (https://osf.io/tz6e7/).

CHAPTER 8 EXERCISES

Exercise 8.1: Finding Transparency

Complete an online treasure hunt to find different types of open science resources and products.

- Read a registered report.
- Obtain video about preregistration.
- Read a manuscript published with all three Open Science Badges.
- Find a job ad in which open science skills are valued explicitly.
- Read an open science blog.
- Download data and materials associated with a published study and reproduce the analyses.

Exercise 8.2: Understanding Manuscript Types: Preregistered Report Writing Instructions

- Download and read instructions from Comprehensive Results in Social Psychology—CRiSP instructions for preregistered reports—www.tandf.co.uk//journals/authors/rrsp-submission-guidelines.pdf.
- List the content and sections that should be prepared for stage 1 submission.
- Describe the role of the editor in the process.
- How is this process different from writing a class paper?
- How is this process different from a traditional paper submission?
- What types of research best match this process?

Exercise 8.3: Online Impact: Reviewing Student Project OSF Page to Avoid Jargon and Initialism

- Review OSF Titles, Description, Wiki Content.
- Eliminate abbreviations, initials, and jargon.
- Share with a peer or family member not familiar with research for extra check.
- Review research paper following same procedures to improve communicability.

END-OF-CHAPTER REVIEW QUESTIONS

1. What makes the Open Science Framework weblinks unique and useful?
2. List three organizations that support scientific transparency in research.
3. Name two examples of citizen science.
4. What can be done with past undergraduate projects?
5. Name three open access sites to help with collaborative research.

6. What is human short-term memory storage capacity?
7. How do acronyms and initialisms violate the ideals of diversity, justice, and sustainability?
8. Name two organizations that promote open science and explain what their missions are.
9. What are the benefits of having undergraduate students join in on crowdsourced projects?
10. Name two crowdsourcing projects and explain what they do.
11. How do free open science tools promote diversity, justice, and sustainability in research?

References

Alogna, V. K., Attaya, M. K., Aucoin, P., Bahník, Š., Birch, S., Birt, A. R., . . . Zwaan, R. A. (2014). Registered replication report: Schooler and engstler-schooler (1990). *Perspectives on Psychological Science, 9*(5), 556–578.

Bench, S. W., Rivera, G. N., Schlegel, R. J., Hicks, J. A., & Lench, H. C. (2017). Does expertise matter in replication? An examination of the reproducibility project: Psychology. *Journal of Experimental Social Psychology, 68*, 181–184.

Byers-Heinlein, K., Bergmann, C., Davies, C., Frank, M. C., Hamlin, J. K., Kline, M., . . . Soderstrom, M. (2020). Building a collaborative psychological science: Lessons learned from manybabies 1. *Canadian Psychology/Psychologie Canadienne, 61*(4), 349.

Chartier, C. R. (2017). *The psychological science accelerator: A distributed laboratory network.* Retrieved from osf.io/93qpg

Citsci_admin. (n.d.). *Home page*. Retrieved February 5, 2021, from www.birds.cornell.edu/citizenscience

Ebersole, C. R., Andrighetto, L., Casini, E., Chiorri, C., Dalla Rosa, A., Domaneschi, F., . . . Vianello, M. (2020a). Many labs 5: Registered replication of Payne, Burkley, and Stokes (2008), study 4. *Advances in Methods and Practices in Psychological Science, 3*(3), 387–393.

Ebersole, C. R., Atherton, O. E., Belanger, A. L., Skulborstad, H. M., Allen, J. M., Banks, J. B., . . . Nosek, B. A. (2016). Many labs 3: Evaluating participant pool quality across the academic semester via replication. *Journal of Experimental Social Psychology, 67*, 68–82.

Ebersole, C. R., Mathur, M. B., Baranski, E., Bart-Plange, D. J., Buttrick, N. R., Chartier, C. R., . . . Szecsi, P. (2020b). Many labs 5: Testing pre-data-collection peer review as an intervention to increase replicability. *Advances in Methods and Practices in Psychological Science, 3*(3), 309–331.

Etz, A., & Vandekerckhove, J. (2016). A Bayesian perspective on the reproducibility project: Psychology. *PLoS One, 11*(2), e0149794.

Gernsbacher, M. A. (2013). *Improving scholarly communication: An online course*. Madison: University of Wisconsin.

Gilbert, D. T., King, G., Pettigrew, S., & Wilson, T. D. (2016). Comment on "estimating the reproducibility of psychological science." *Science, 351*(6277), 1037–1037.

Global Garlic Mustard Field Survey. (n.d.). Retrieved February 5, 2021, from www.garlicmustard.org/

Grahe, J. E. (2010). *Collective undergraduate research project*. Retrieved April 28, 2021, from https://sites.google.com/a/plu.edu/curp/

Grahe, J. E. (2014). Announcing open science badges and reaching for the sky. *Journal of Social Psychology, 154*, 1–3.

Grahe, J. E., Brandt, M., IJzerman, H., & Cohoon, J. (2014). Replication education. *APS Observer, 27*(3).

Grahe, J. E., Chalk, H. M., Alvarez, L. D. C., Faas, C. S., Hermann, A. D., & McFall, J. P. (2018). Emerging adulthood measured at multiple institutions 2: The data. *Journal of Open Psychology Data, 6*(1).

Grahe, J. E., Reifman, A., Hermann, A. D., Walker, M., Oleson, K. C., Nario-Redmond, M., & Wiebe, R. P. (2012). Harnessing the undiscovered resource of student research projects. *Perspectives on Psychological Science, 7*(6), 605–607.

Greenberg, J., Pyszczynski, T., Solomon, S., Simon, L., & Breus, M. (1994). Role of consciousness and accessibility of death-related thoughts in mortality salience effects. *Journal of Personality and Social Psychology, 67*(4), 627–637.

Hales, A. H., Williams, K. D., & Rector, J. (2017). Alienating the audience: How abbreviations hamper scientific communication. *APS Observer, 30*(2).

Kathawalla, U. K., Silverstein, P., & Syed, M. (2021). Easing into open science: A guide for graduate students and their advisors. *Collabra: Psychology, 7*(1).

Klein, R. A., Cook, C. L., Ebersole, C. R., Vitiello, C., Nosek, B. A., Chartier, C. R., . . . Ratliff, K. (2019). *Many labs 4: Failure to replicate mortality salience effect with and without original author involvement*. Retrieved from https://psyarxiv.com/vef2c/

Klein, R. A., Ratliff, K. A., Vianello, M., Adams, R. B., Bahník, Š., Bernstein, M. J., . . . Nosek, B. A. (2014). Investigating variation in replicability. *Social Psychology, 45*(3), 142–152.

Klein, R. A., Vianello, M., Hasselman, F., Adams, B. G., Adams, R. B., Alper, S., . . . Sowden, W. (2018). Many labs 2: Investigating variation in replicability across samples and settings. *Advances in Methods and Practices in Psychological Science, 1*(4), 443–490.

Kleinberg, B., Nahari, G., Arntz, A., & Verschuere, B. (2017). An investigation on the detectability of deceptive intent about flying through verbal deception detection. *Collabra: Psychology, 3*(1).

Moshontz, H., Campbell, L., Ebersole, C. R., IJzerman, H., Urry, H. L., Forscher, P. S., . . . Chartier, C. R. (2018). The psychological science accelerator: Advancing psychology through a distributed collaborative network. *Advances in Methods and Practices in Psychological Science, 1*(4), 501–515.

National Park Service. (2020, November 24). *Work with us*. Retrieved February 12, 2021, from www.nps.gov/subjects/youthprograms/jobs-and-internships.htm#CP_JUMP_5459385

Nosek, B. A., & Bar-Anan, Y. (2012). Scientific utopia: I. Opening scientific communication. *Psychological Inquiry, 23*(3), 217–243.

Nosek, B. A., Spies, J. R., & Motyl, M. (2012). Scientific utopia: II. Restructuring incentives and practices to promote truth over publishability. *Perspectives on Psychological Science*, *7*(6), 615–631.

Open Science Collaboration. (2012). An open, large-scale, collaborative effort to estimate the reproducibility of psychological science. *Perspectives on Psychological Science*, *7*(6), 657–660.

Open Science Collaboration. (2015). Estimating the reproducibility of psychological science. *Science*, *349*(6251).

Rissman, L., & Jacobs, C. (2020). Responding to the climate crisis: The importance of virtual conferencing post-pandemic. *Collabra: Psychology*, *6*(1).

Rosenthal, R. (1979). The file drawer problem and tolerance for null results. *Psychological Bulletin*, *86*(3), 638.

Sætrevik, B. (2021). Realistic expectations and prosocial behavioural intentions to the early phase of the COVID-19 pandemic in the Norwegian population. *Collabra: Psychology*, *7*(1).

Sassenhagen, J., Blything, R., Lieven, E. V., Ambridge, B., Zwaan, R., & Ferreira, F. (2018). Frequency sensitivity of neural responses to English verb argument structure violations. *Collabra: Psychology*, *4*(1).

School Spirit Study Group. (2004). Measuring school spirit: A national teaching exercise. *Teaching of Psychology*, *31*(1), 18–21.

Schweinsberg, M., Madan, N., Vianello, M., Sommer, S. A., Jordan, J., Tierney, W., . . . Uhlmann, E. L. (2016). The pipeline project: Pre-publication independent replications of a single laboratory's research pipeline. *Journal of Experimental Social Psychology*, *66*, 55–67.

Science united. (n.d.). Retrieved February 5, 2021, from https://scienceunited.org/

Silberzahn, R., Uhlmann, E. L., Martin, D. P., Anselmi, P., Aust, F., Awtrey, E., . . . Nosek, B. A. (2018). Many analysts, one data set: Making transparent how variations in analytic choices affect results. *Advances in Methods and Practices in Psychological Science*, *1*(3), 337–356.

Simons, D. J., Holcombe, A. O., & Spellman, B. A. (2014). An introduction to registered replication reports at perspectives on psychological science. *Perspectives on Psychological Science*, *9*(5), 552–555.

Spellman, B. A. (2012). Introduction to the special section: Data, data, everywhere . . . especially in my file drawer. *Perspectives on Psychological Science*, *7*(1), 58.

Steegen, S., Tuerlinckx, F., Gelman, A., & Vanpaemel, W. (2016). Increasing transparency through a multiverse analysis. *Perspectives on Psychological Science*, *11*(5), 702–712.

Stroebe, W. (2019). What can we learn from many labs replications? *Basic and Applied Social Psychology*, *41*(2), 91–103.

Uhlmann, E. L., Ebersole, C. R., Chartier, C. R., Errington, T. M., Kidwell, M. C., Lai, C. K., . . . Nosek, B. A. (2019). Scientific Utopia III: Crowdsourcing science. *Perspectives on Psychological Science*, *14*(5), 711–733.

Van Bavel, J. J., Mende-Siedlecki, P., Brady, W. J., & Reinero, D. A. (2016). Contextual sensitivity in scientific reproducibility. *Proceedings of the National Academy of Sciences*, *113*(23), 6454–6459. doi:10.1073/pnas.1521897113

Wagge, J. R., Brandt, M. J., Lazarevic, L. B., Legate, N., Christopherson, C., Wiggins, B., & Grahe, J. E. (2019). Publishing research with undergraduate students via replication work: The collaborative replications and education project. *Frontiers in Psychology*, *10*, 247.

Welcome to journey north. (n.d.). Retrieved February 5, 2021, from https://journeynorth.org/

What is seti@home? (n.d.). Retrieved February 5, 2021, from https://setiathome.berkeley.edu/

PROGRESS

Open Science Promotes Diverse, Just, and Sustainable Outcomes

Chapter 9 Objectives

- What has changed (or not) about racist attitudes
- Consider how diversity, justice, and sustainability can fit into everyday life and research
- Recognize that research is done on a homogenous population
- Define WEIRD researchers and participants
- Explore diversity in research
- Explore social justice in research
- Explore sustainability in research
- Examine all of diversity, justice, and sustainability together
- Discuss how open science benefits diversity, justice, and sustainability

MUSIC IN A BOX: ABOUT THE SONG "PROGRESS: OPEN SCIENCE PROMOTES DIVERSE, JUST, AND SUSTAINABLE OUTCOMES"

One of the more difficult songs to create, this one is more contemplative and musically more challenging and complex. The lyrics to this song are a painful reminder of the legacy of discriminatory science practices against persons of color as well as women and bluntly call attention to the importance of social issues (such as racial, ethnic, gender, and sexual identities) in scientific practice. The lyrics also remind the listener that open science offers solutions through better science that reduces our personal biases through greater diversity with more just and sustainable processes. The song suggests that open science initiatives provide some platforms to achieve these goals.

WHAT HAS CHANGED?

At one point during my Crisis Schmeisis tour, I stayed on a couch as a guest in a 200+ year-old farmhouse in Maryland. My host took me on a tour of the house and property including a tiny shack that originally housed enslaved people. I was taken aback that it was still standing, but the building was still solid with a lock on the outside but not the inside and the metal rings where chains were threaded to keep enslaved people in shackles. I stood on the dirt floor and looked at the stone walls with no windows. As the day progressed and I looked around the landscape, I pondered how much had really changed in the 150 years since that building was used for enslaved people. As we prepared dinner, the building loomed in my view, as did the pastoral scenery on the hillside behind it. I asked my host to ponder his thoughts about the changes in the last 150 years. His responses were disappointing, both in their lack of empathy for the people who suffered at the hands of white supremacy then or now and in the jovial approach he took to the situation.

The next day, I toured Antietam Battlefield, part of the National Park Service. On September 17, 1862, a combined total of 22,717 Americans died fighting on both sides of the Civil War. It was the bloodiest single battle of the war and in US history, and it highlights to me one of the worst days in US history in valuing diversity, social justice, and sustainability. From my own perspective, here is why. Enslavement was historically a diverse trade in which any victims of aggressive empires might be sold into a life of enslavement. Romans, for instance, had no color preference for enslaved people. However, in the American colonies in 1619, enslavement codified the roles of enslaved people into people of color, particularly Black people. Because only certain people were considered slave stock, the process violates diversity. Because enslavement is an immoral act in which people treat others as property, the trade is socially unjust. Because the Civil War was fought directly and indirectly over whether to end enslavement, it is a social justice battle intended to improve the valuing of diversity. However, Northerners were just as racist as Southerners, even those who were abolitionists (Aptheker, 1991; Blight, 1999). The lost life associated with war, including not only humans but also animals and plant life, is an affront to efforts to create a sustainable society.

As I toured the many miles of the Antietam Battlefield, mourning the loss of human life, I thought about how many crops were destroyed that day as the men crossed through planted fields to kill each other. I considered the destruction of buildings, roads, rail lines and the damage to natural ecosystems along the river banks and in the forests. Bullets and cannonballs do not distinguish between targets; they only destroy objects in their path.

Of course, this was one day 165 years ago, and weapons of war and the sizes of armies have only grown larger and more damaging in the time since. I find it easy to see how war and conflict is the opposite of valuing diversity, social justice, and sustainability. However, the other end of the continuum is harder to envision, and it continues to elude us, demonstrating its challenge to achieve. There is much evidence that the field of psychology (along with other social sciences) is often focused on advancing research questions aimed at increasing the individual and societal values placed on diversity, social justice, and sustainability. However, there is no one clear path to any of these three constructs, as they can lead to conflict with each other when simultaneously advanced. In this chapter, I will first consider each of these individually and how they relate to open science research. Finally, I will consider them interactively to show their potential conflict and the possible compromises that help them work together.

CONSIDERING DIVERSITY, JUSTICE, AND SUSTAINABILITY

My example for diversity was regarding color, specifically as it related to race and enslavement. Though the problems of slavery and race continue to be persistent and critical issues within the United States, the concept of diversity is much broader. Who are you? In what ways are you like most humans? How do you differ? Table 9.1 shows my personal responses to this task.

DIVERSITY IS COMPLEX

By creating both the categories and the answers, the readers can see what part of my identity is most important and how I respond to commonly considered categories. It is certainly not a complete list, as is recognizable after doing the task. There are categories that I do not feel comfortable sharing, as they are too personal, and there are too many categories to consider in one location. However, I tried to demonstrate that I recognize my whiteness and my maleness as parts of my identity. Because of who I am, there are some aspects of social injustice, exclusion, and depredations I can only imagine. Because of who I am, there are many psychological questions and phenomena that have no interest to me. Because of who I am, there are aspects of the research process in

Table 9.1 Similarities and Differences from Others.

Characteristic	*Same as Others*	*Different Than Others*
DNA	99.999% shared	Configurations of DNA
Food, water, shelter	Needs	Preferences
Motivations	Strive to match values	Relative importance of values
Age	Ranges till death	Still alive, middle-aged
Gender	Identity matters	Heterosexual male
Race	Color shouldn't matter	My whiteness carries privilege
Socioeconomic status	Impacts all facets of life	Middle class from privilege/hard work
Mental health	Predisposed to disorder	Predisposed toward bipolar/suicide
Family network	Mostly provides support	Disadvantaged due to location/context
Social networks	Mostly provide support	These feel unique across humanity.
Hobbies	Critical to self-development	Hiking, music, billiards are rewarding

which I excel but others in which I struggle or even fail. Because of this, I can only answer some questions, and fewer questions well.

Who Matters in Research?

Bluntly stating my own strengths and weaknesses is intended to highlight the importance of diversity of the researcher. For too much of the history of science in Western cultures, most groups were discouraged, if not actively excluded, from the education needed to be researchers or the jobs even if qualified. The first woman of color to earn a Ph.D. in the US was Dr. Arai Haraguchi, in 1905, a Japanese woman who came to the US to earn a degree. In a project dedicated to collecting dissertations from women of color who earned Ph.D.s before 1970, we struggled to find their names. And even those who were found often experienced unfair challenges impeding their progress or ability to achieve success. While the explicit limitation of certain people from achieving success is a question of social justice, the impact is that there has been a disproportionate number of white male researchers in academia (Saini, 2017; Cynkar, 2007). Though recent education and trends suggest this will change in time, particularly in psychology; at present, the next generation of researchers is still being trained by people who do not share the same personal identity as they do.

This bias toward white people, and men in particular, impacts other aspects of the research process because the publication process is based on peer review. Thus, the reviewers and editors who decide what gets published are disproportionately white men. Ignoring social justice concerns, of which there are many, the lack of diversity makes it likely that some topics will not be considered important. Further, while reviewers should be blind to author identity, still today, editors are not blind to author identity. Therefore, any implicit biases inherent in the editor could be primed while reviewing author information during manuscript processing. This is not solely a problem of white maleness, because psychology researchers also tend to be ideologically liberal. Haidt and Jussim (2016) and others (Redding, 2001; Duarte et al., 2015) perceive and criticize a liberal bias against conservative ideas in psychological research.

What Is WEIRD?

Henrich, Heine, and Norenzayan (2010) first labeled the phenomenon whereby research participants in psychology tend to be Western, educated, industrialized, rich, and democratic (WEIRD), even though that is a minority of the worldwide population. And much like the word "weird," WEIRD populations are unusual. If the majority of the world is not Western, highly educated, industrial, rich, or democratic, then why are they studied more than anyone else? The simple answer is that psychological researchers are also WEIRD, and as presented in the preceding paragraph, they also overrepresent white people and men.

It is mansplaining to even point out that this is a problem. It is unrealistic to expect scientists who represent 12% of the total population (Azar, 2010) to fully represent the questions of the many billions of other people on the planet. Non-WEIRD research is on the rise globally in Asia, South America, and Africa as access to research and education increases. Or perhaps it would be better to say that access rises in non-WEIRD cultures as barriers to research are diminished. When both the researchers and the participants come from WEIRD cultures, other cultures are invisible in the process altogether (Syed, Santos, Yoo, & Juang, 2018). This creates a disconnect between health services and nondiscrimination, with mental health demonstrating the largest effect (Carter, Lau, Johnson, & Kirkinis, 2017).

Even as we focus on a more global population to study, it strikes me that even WEIRD people get older. The same factors that lead to a preponderance of WEIRD researchers, participants, and questions also lead to a preponderance of research being conducted on 18- to 22-year-old college students. While the advent of M-turk and other resources that involve other adult populations helps, we still make too many assumptions about the generalizability of research targeting college students as relevant across the ages. For that matter, as a researcher who abandoned the topic of studying groups due to lack of resources, there are certain questions that are not explored because they are too difficult or too costly to measure effectively. More importantly, studying topics with intensive resource costs can make it more difficult to get a job. All of these limits to diversity constrain the generalizability of research findings.

Diversity Is More Than Skin Deep

Race does not predict any systematic variance beyond the sociocultural factors that amplify the categorization. In other words, skin color does not predict psychological variables after taking into account social and systematic factors. For a blunt concrete example, any differences between "races" on IQ scores are better explained by sociocultural than genetic factors, despite a long history of psychology attempting to characterize intelligence differences across racial groups (e.g., Herrnstein & Murray, 1994) and as catalogued by Guthrie (2004) in *Even the Rat Was White*. Rather than race, culture and ethnicity explain much more about a person and their identity. Cultural factors such as education and religion are often correlated with race to some degree, but race encompasses a few arbitrarily assigned skin colors, whereas people with each skin color represent many populations who differ along underlying characteristics. There are 54 countries in Africa, 48 in Asia, and 14 in South America. Within each country, there are minority groups and regions that also contribute to cultural differences. Anyone who has traveled across various parts of the US knows that countries include subcultures. Multiplied across the hundreds of countries and regions, the concept of cultural diversity is recognizably complex.

Another diversity category that is more complex than previously considered is that of biological sex and gender. As was standard practice when I was being trained, my demographic measures included two boxes for biological sex, one for male and one for female. However, in the modern age, science recognizes this as a false dichotomy. While psychology recognized gender as continuous, the field persisted to measure sex as a dichotomy. The transgender community has demanded change. It is not enough to consider gender as dichotomous, but rather to acknowledge that it is continuous. Now, to measure biological sex, I build surveys with an open-ended text box so that respondents are not constrained by a false dichotomy or insulted by having to check "other" before reporting. While the majority of responses are still male and female, the survey is now inclusive and values diversity.

Under most circumstances, we are not biologically driven to act because of our skin color or our biological sex. However, culture and society may influence us to act in certain ways because of our skin color or sex. Rather, we act to achieve goals and outcomes that match our desires and values. Even under circumstances in which choices are constrained by lack of resources or restrictive laws, our actions are guided by our strivings for life.

Here is where diversity gets incredibly complex. That second research question I ever tried to answer as a senior in college was to examine how religious affiliation associated with Schwartz's (1992) basic values. My findings are not worth considering because they were never peer reviewed, but the question helps imagine complex diversity of values and beliefs. Schwartz's approach to values is still useful today, and his overview (Schwartz, 2012) shows a stable structure of those 10 basic values still today (self-directions, stimulations, hedonism, achievement, power, security, conformity, tradition, benevolence, and universalism). Aligned on two axes (Openness/Resistance to change and Self-Enhancement/Transcendence), people are unlikely to share opposing values. For instance, correlations between hedonism and conformity or universalism and power are negative. Additionally, the values that are situated orthogonally share no correlations. In other words, someone who values Achievement could as easily be driven by discovery in valuing Stimulation and Self-Direction as they are by conserving history through maintaining Tradition through Conformity and Security. Trying to explain interaction terms in a results section becomes increasingly complex with additional explanatory factors. Ten values on two axes are complex, but that is compounded when considered in the context of religion.

There are 12 major and countless minor religions. In my college project, I examined four different denominations out of the dozens of one of the major religions (Christianity). My population was WEIRD because that is what I had access to, but to explain how religion might predict human behavior, all major and minor religions should be considered. And yet religion itself is not a sufficient construct, because people differ in how they engage in religious practices and how they relate to the institutional beliefs and assumptions.

Further, contexts under which the religion is practiced impact how those beliefs are manifested. For example, does the culture value or restrict religious beliefs, and are cultural values codified into law where practices might lead to penalties and imprisonment?

As white people settled across North America, they often made laws that limited or eliminated indigenous cultural practices. This included actively retraining children into Christianity to save them from their heretical native beliefs. Readers might take a moment to imagine a missionary administering severe corporal punishment to one or more of their pupils to dissuade them from using their native language, or the tears of those children. Now amplify that pain to represent the suffering across a continent including torture, rape, and murder until people were subjugated and relegated to reservations, where their behavior can be limited and lives constrained. Often, those actions were justified as supporting a chosen people to settle a country they received from their own god, a concept known as Manifest Destiny.

Sometimes, I try to imagine the psychology of people living through these experiences. When I see images of similar actions happening now across the world, I wonder about the similarities and differences as I recoil in horror while being thankful for my own privileged security. To explain psychology in the modern age, the model certainly needs to consider more than basic values and religion; the motivations and experiences that guide us are uniquely complex, leading to universal complexity.

Social Justice in the Context of Research

After some discussion of diversity, we now turn to consider how those diverse people are treated. Social justice is the "relation of balance between individuals and society measured by comparing the distribution of wealth differences, from personality liberties to fair privilege opportunities," as defined by Wikipedia. When considered from a social justice perspective, the settling of the United States was incredibly unjust, and that is an understatement. It was the systematic murder and subjugation of many groups of people to advance the goals of another group of people. Consider further that legal enslavement existed for much of this period, meaning that some people were forced to subjugate others, if only in the role of supporting aggressors' wealth. After enslavement was outlawed, US history is littered with periods in which individuals supporting totalitarian, fascist agendas supporting white supremacy garner sufficient power to make laws limiting the rights of others, often targeting African Americans and Native Peoples, but there are ample examples of subjugation of all people of color, in every state in the United States, not just the South. Limits on where they could live, the jobs they could hold, the education they could earn, and the wealth they could accumulate were often compounded by sporadic acts of rage and rioting that increased fear and destroyed progress made by communities of color.

Psychological research tries to speak to this recurring threat of totalitarian agendas. Personality research identified a "dark triad" of traits—narcissism, Machiavellianism, and psychopathy (Furnham, Richards, & Paulhus, 2013)—that might lead to personality types that support these value systems. Social psychology identified that both motivation and ability to process a message fully are required in order to recognize its true meaning (Petty & Cacioppo,

1984) and that interfering with either process makes it easier to persuade belief with a bad argument. More recently, research has focused on the proliferation of hate speech and recruiting of new followers to these extremist groups (Cohen-Almagor, 2018; Darmstadt, 2019). These are only a few examples of questions posed by psychology researchers, and they demonstrate my bias from training as a social and personality psychology researcher. Biological, cognitive, developmental . . . frankly, the whole alphabet of psychology subdisciplines address questions regarding how to make society more socially just. As evidence, remember the criticisms that psychological research is too liberal.

However, psychology also has a history of engaging in and supporting research that was used to subjugate others. The most startling example might be one of the most important contributors to the field, Sigmund Freud. Whether intentional or not, his theories regarding sexuality were derogatory toward women and used to justify and advance draconian morality regarding sexual interactions for much of the 20th century. Personality psychologists developed methods and theories that favored their own construction of the world (predominately white, male, and Christian), which led to theories that certain races were more intelligent than others (Brigham & Yerkes, 1923; Goodenough, 1926) and more capable of certain doing jobs (Woodworth, 1916; Garth, 1934). Because women were subjugated through voter suppression and other cultural systems, psychology research is often driven by men, and many examples of early research are demeaning to women when considered with modern understanding. Other aspects of sexuality were also oppressed through the guise of authority. Until 1973, homosexuality was listed as a psychological disorder (Bayer, 1987), and the treatments included medical, cognitive, and physical approaches to convince the victim to rebel against their own sexuality.

Even if modern psychology is driven toward a more socially just world, the scars of a century of oppressive research still resonate today. In the summer of 2020, the US population seemed finally able to recognize the need for conversations about social justice. However, the continued presence of systematic violence against people of color, particularly against African Americans, as well as the challenge in conveying the underlying causes of that violence and suppression, demonstrate that research evidence does not translate to better practice. This is compelling evidence that our field and other social sciences need to amplify our concern for social justice (Cook, 1990; Lerner, 1977), not only in the questions that are asked (Darwin, 1919) but also in the systems that advance those questions (Fassinger & Morrow, 2013).

A later section of this chapter is focused on connecting open science initiatives to these advances in social justice. To appreciate their potential in alleviating oppression and advancing equality, some evidence of the barriers related to equality in psychological research and psychological barriers to equality should be reviewed. But first, the distinction between equality and equity must be made. Equality reflects that all persons should have the same opportunities and rewards, whereas equity is the process of repairing past inequalities in order to make a future that is equal. For example, the procedures commonly known as affirmative action allowed African American communities to engage in educational and economic opportunities that were previously unavailable to them. Scholarships that benefit women or people of color are equitable in offering these groups extra opportunity because they were historically excluded from the same opportunity. When people of power then complain that they do not have enough opportunity, it reflects the loss of privilege.

Biology Breeds Bias

The psychology of stereotype, prejudice, and discrimination that underlies the processes that lead to inequality is deeply ingrained in our psyche from deep in our evolution. All my dogs would love to catch and eat the mice, rats, squirrels, or raccoons that pass through our backyard, but they tolerate (affectionately?) our cats. This is because they learn to categorize friends and enemies. Barking at the knocking at a door yields either continued barking or tail wagging and panting depending on the person who enters the doorway. It is not the dog's biology that helps them distinguish animal and human friends but rather that their biology allows them to categorize others as friends or enemies.

Humans are remarkably more complex than dogs, but this basic predisposition to categorize others remains. Stereotypes are generalizations about people in categories, prejudices are the biases and preferences for people based on those categories, and discriminations are acts that are directed toward people because of those stereotypes and prejudices. Alternatively, stereotypes are cognitive, prejudices are affective, and discriminations are behavioral responses toward others based on categories. Predominantly, these are negative, but they can be positive as well. So my example shows that dogs demonstrate stereotyping, prejudice, and discrimination by preferring certain others and disliking or even attacking different others. Humans do the same thing, but we have the capacity to recognize when we are taught something that is wrong. Dogs always trust their masters.

While we have the capacity to unlearn the justification for stereotyping, prejudice, and discrimination, there is little evidence that we can unlearn those initial connections once learned. "Implicit bias" is a term used to describe unknown and unintended prejudices we accumulate across time. Project Implicit (www.implicit.harvard.edu/) is a location where you can measure some of your own implicit biases. Of course, these biases reflect internal and sometimes unknown prejudice and are only marginally predictive of discriminatory acts. However, the existence of bias preferences to others, implicit or explicit, is a threat to equality because in-group bias is a universal phenomenon. Even when formed using random assignment, groups start viewing outgroups as more homogeneous and fellow in-group members in more positive ways. In a laboratory, giving a few extra dollars to random strangers arbitrarily bound to us is one thing, but in practice, it leads to systematic injustice.

Earlier, I pointed out that editors are never blind to author identity. Consider that fact in light of implicit bias and in-group favoritism. There are two ways in which editors influence the outcome of a manuscript in their queue: the reviewers they select and their interpretation of the research. Their reviewers are often drawn from a pool who regularly review. Here, implicit bias could lead to matching in-group members' research with an easier reviewer or reviewers in favor of the theories being advanced. The practice of rewarding in-group members and discriminating against out-group members is well-established. So is the fact that in-group bias effects emerge based solely on names (Bertrand & Mullainathan, 2004) or other characteristics such as educational or cultural background even if the two individuals have never met. Assuredly, editors would claim that such bias is beneath them, but animal researchers said the same thing of expectancy effects; they were wrong (Rosenthal & Jacobson, 1968). No human is above their implicit biases, and therefore, systems should be in place to minimize bias whenever possible. As will be offered soon, open science has proposals to reduce some of that systematic bias in research and publishing.

Sustaining Progress Is Limited by Definitions

Imagine the ideal future. What would be an ideal future for humanity? Does it include relative equality across all people who get to choose their own mechanisms to find meaning and purpose in life while living harmoniously with natural ecology in pastoral settings? Or does it include a society driven toward the stars finding new forms of energy and solutions to any impediments toward a species that dominates a solar system? Maybe your future is radically different from both these options. Unless we are living an ideal existence now, there is a discrepancy between the current state of the world and a utopian future. Movement toward the ideal from the present would represent progress, and systems that maintain that progress across time and space would enable sustainable progress.

This highlights the core problem for any social justice movement: defining progress and the mechanisms to achieve those goals will illuminate inherent challenges between competing expectations for what represents progress. The two ideal futures imagined pit two types of progress against each other. To achieve an ideal world of harmony between humans and the natural systems of the planet, society needs to make sacrifices in ways that we produce food and energy and how we build our homes and cities. The costs of these sacrifices conflict with the ideal world filled with urbanized space travelers. A society intent on conquering space would seek more sources of energy regardless of the cost to natural systems, whereas harmonious pastoralists would value the systems that protect biological diversity over industrial development.

My extreme examples make the distinction of competitive goals toward progress clearer to spot, but life is more complex. When advancing to social justice, it is beneficial to consider the ideal future that is desired. Perhaps we would do better to explicitly convey our ideal future in conversations about how to resolve injustice. Differences emerge when developing systems trying to resolve conflicts in race relations such as systematic violence against Black people, disenfranchisement of minority groups by majority society, reducing negative impacts of drug use, prostitution, and domestic abuse.

Assumptions about causality and attributions of blame guide interventions to target individuals versus systems. Someone who believes anyone can achieve anything if they just set their mind to it will create a very different system than someone who believes that institutions create barriers that prohibit individuals from achieving their desired goals. Even those with shared goals can still disagree on selecting ideal solutions or identifying which strategies would achieve them. Even a goal as simple as reducing criminal activity becomes complicated, because people have different opinions on how this effect should be achieved. Some will want to create harsher punishments, thinking that this is the way to deter someone from committing a crime. Others feel that decriminalizing certain acts and creating better support systems will decrease the motivation for crime.

Sustainability in the Context of Research

The question of sustainability in research programs extends to any person or program dedicated to advancing science in the sense that humans need to be sustained. Maybe it seems silly to consider, but researchers need to eat, drink, sleep, and find shelter. Psychological needs also need to be met for people to be productive. They need to feel safe, have time to think, and possess self-efficacy to conduct the research. Can you imagine a situation in which these basic needs are missing from the lives of professional researchers? I certainly can. After decades of waning governmental support for education and general challenges in the traditional economic model for higher education, there are more Ph.D.s on the market looking for jobs than there are jobs for them to earn. After focusing on undergraduate and graduate school for 5 to 10 or even more years, Ph.D.s often carry large amounts of student loan debt while being paid entry-level wages that might not cover the cost of living.

Researchers who do find work and stable incomes still face sustainability challenges including maintaining sufficient productivity to keep the job. Like other institutions, colleges and universities vary in their inherent resources. Psychology researchers need access to equipment and materials in addition to human participants. This resource challenge is partly to blame for the reliance on college-aged WEIRD populations, because they are available and cheap. However, other costs creating obstacles include access to library resources and technology. The cost of journal subscriptions continues to rise, and libraries are struggling to keep up. Additionally, the cost of technology is ongoing, and stretched budgets might include cuts to replacement computers, software, or other laboratory materials. Finally, the costs of labor and travel associated with research extend beyond the salary of a faculty member. Research assistants and graduate students require their own salaries, and travel and professional membership costs are incurred along the way.

Individuals with greater personal wealth or who serve wealthier institutions are advantaged with greater sustainability than those who struggle with resources. If these individuals all had equal access to those resources, then the merit system would be rewarding excellence, but we know that socially unjust systems historically limited wealth accumulation of persons of color as well as their access to prestigious and wealthy institutions.

Limits to diversity and socially unjust systems lead to questionable outcomes. One concept representing this impact is undone research (Bhambra, Gebrial, & Nişancıoğlu, 2018) or work completed by persons representing disadvantaged groups that was missing because capable individuals never got a chance to research. Until 1940, few Black women earned Ph.D.s in psychology (three were Inez Beverly Prosser, Ruth Howard, and Alberta Banner Turner). They were the first to earn this degree and are rare examples in the field. Their work is rarely represented in any textbooks beyond noting their accomplishment. The lack of representation reflects not that fewer people were capable of earning that degree but rather that educational opportunities were discouraged or explicitly blocked. Further, even with a degree, sex roles limited opportunities. Mamie Phipps Clark serves as an example for this in the mid-20th century. Though she was the primary researcher of the groundbreaking studies used as evidence in *Brown v. Board of Education*, which served to desegregate American schools, her husband, Kenneth, is frequently cited in history-of-psychology texts as the lead researcher or given equal credit (Cramblet Alvarez, Jones, Leach, & Rodriguez, 2020). Moreover, she felt the impacts of her intersecting identities as she attempted to gain employment following her Ph.D., famously noting:

> although my husband had earlier secured a teaching position at the City College of New York, following my graduation it soon became apparent to me that a Black female with a PhD in psychology was an unwanted anomaly in New York City in the early 1940s.
>
> (Phipps Clark as cited in O'Connell & Russo, 1983, p. 271)

Further, the question of how to sustain any system should be an active component of any research program. For instance, systems designed to reduce systematic racism should consider not only methods to implement change but also how to maintain those systems across an extended period. Any proposed intervention will experience challenges from opposition who are either threatened by change or because of fear encouraged by a racist agenda. In the United States, that opposition is significant enough to threaten any real advances in social justice. If interventions that worked were also ensconced in sustainable models, the world would already be better than it is. Methods for building more inclusive institutions are demonstrated on small scales or for periods of time. And yet metrics suggest little measurable change in the education or wealth gaps between white people and other groups. When extended to gender, women still earn less than men when in the same jobs and are still underrepresented in upper management and leadership positions.

In conclusion, research is less generalizable and sustainable when large portions of the population are excluded from the process. Research that intends to address these social injustices needs to include models for sustained application, or perhaps a new field of research should emerge that measures how to extend and increase the impact of effective social justice interventions.

Of course, if humans continue to disregard our consumption relative to resource rejuvenation, perhaps the field of psychology will no longer be relevant at all, because without a healthy planet on which to live, there are threats to our very existence. Science on the impacts of humans on the environment proffered warnings about greenhouse gasses and climate threats for many decades. There are many challenges in resolving humanity's unchecked population growth and subsequent resource consumption. These include issues such as (a) convincing people to sacrifice comfort or perceived comfort, (b) creating credible counterarguments against religious ideologies that do not value natural systems, (c) unseating greedy and powerful oligarchs and oligopolies that control most of the world's resources, (d) unifying world governments around a common goal that simultaneously upends historical power structures. There are meaningful roles for psychology and other social sciences to provide knowledge about how to address these challenges. In sum, humanity needs sustainable solutions to environmental questions in addition to existing concerns.

CONSIDERING DIVERSITY, JUSTICE, AND SUSTAINABILITY TOGETHER

Since the summer of 2020, as the Black Lives Matter protest movement grew in massive proportions, a proliferation of diversity, equity, and inclusion (DEI) programs has emerged. Major sports organizations such as the NFL enacted campaigns promoting DEI, in stark contrast to their initial response when Colin Kaepernick famously took a knee during the National Anthem to protest institutional violence against Black people. At one level, I am pleased to see a conversation about social inequity occurring in ways that has not happened previously in my lifetime. At another level, I am troubled that those concerned with DEI are not more concerned with sustainability as well.

Climate change due to human activity results from energy production in addition to increased water and land consumption. It is not simply a problem of over-population in which 7+ billion people are more than the planet can sustain. Rather, it is a social inequity problem. The WEIRD people consume much more of the relative world resources than those in other parts of the world. In contrast, the impacts of climate change such as land loss to rising waters or inhospitable environments due to excessive temperature or drought are harder on people in poor countries with developing economies (Swim & Bloodhart, 2018). Thus, diversity and social justice questions must consider sustainability at macro levels across local, state, even federal governments to bring forth a world economy that is more diverse and equitable. At the same time, to change these systems, people must engage in that change.

One criticism of environmental movements is that they value animals and plants over people. A thick, impenetrable forest might be critical to help cool the planet across eons, but people want to use those resources now. For people living at or below poverty levels, it is hard to balance care for planetary resources with a job that might help keep family members from suffering. These questions need to ask how we improve the planet while benefiting humanity so that program objectives do not contradict themselves.

HOW DOES OPEN SCIENCE BENEFIT DIVERSITY, SOCIAL JUSTICE, AND SUSTAINABILITY?

In Grahe, Cuccolo, Leighton, and Cramblet Alvarez (2019), we asserted a number of ways that open science initiatives supported and advanced diversity, social justice, and sustainability. By this point in the book, readers might anticipate these already. First, the Replication Crisis manifested from a series of threats to diversity, social justice, or sustainability. For instance, standards allowing for sparse reporting of methods and results details cultivated an environment that fostered questionable research practices. Publication bias preferred flashy, novel, and statistically significant findings over rigorous replications that questioned prior findings. Systems that preferred individual labs following an excellence model limited power to publish to a small cadre that ignored diverse voices warning that findings were not reproducible.

As open science approaches try to increase methodological rigor, various byproducts benefit diversity and social justice. For instance, crowd projects offer opportunities to those previously excluded. As Auchincloss et al. (2014) noted, classroom-based crowd projects directly benefit students from underprivileged backgrounds because they are less likely to be invited or seek involvement in summer or independent study research experiences. At a professional level, these also proffer benefits. Crowd projects often make big impacts as measured by the status of the journal they are published in and the citation rate (impact factor). Finally, they offer networking and leadership opportunities that were previously limited to individuals with prestige.

Other systems similarly yield advances as byproduct rather than intentional effort. The effort to extend the availability of science tools and outcomes through online repositories such as the OSF or preprint servers such as PsyArXiv invite anyone to become their own publisher. If someone pursues research that contradicts the zeitgeist accepted by powerful editors and established reviewers, posting the work in public spaces allows for dissemination. While I have no examples of research that match this description at present, only a short period of time has passed since the possibility has existed. However, it is now true that anyone could conduct research and share it with the entire science world. They would still need to find readers to adopt the assertions presented, but the ability to share the research exists.

Though there are many examples of benefits as byproducts of open science, some organizations are actively noting their intentions to approach research from a more diverse and equitable perspective. The Psychological

Science Accelerator codified its valuing of diversity and social justice in its organizing documents and actively seeks to include individuals from non-WEIRD countries and underrepresented groups as contributors and leadership positions (Moshontz et al., 2018). Similar organizational language was added to the Society for Improving Psychological Science (SIPS), and they created travel grants and leadership positions targeting greater inclusion as well. SIPS is not the only professional organization dedicated to crafting a better science. Every professional organization to which I belong made some effort to show their support social justice in the summer of 2020, and some dedicated resources to enacting change. Of course, if these systems do not consider the sustainability of their interventions, the change could be short term. And yet whether the outcomes are byproducts or intentionally planned, open science initiatives offer many opportunities to advance diverse, socially just, and sustainable research.

Chapter 9: Crisis Schmeisis Journey Book Reviews

Proud Shoes

Pauli Murphy

This book is written by a woman whose family was a combination of two mixed-race families. The discussions of how race was conceptualized reflect the constructed nature of the construct. The changing impact of cultural attitudes about race are evident as she describes her family's challenges and success navigating personal and institutional racism (https://osf.io/tz6e7/).

The Legacy of Luna

Julia Butterfly

This book describes the events experienced by a woman who lived in an old-growth tree for more than a year in an effort to keep it from being cut down. Her story highlights the social injustice associated with clear cutting old-growth forests (https://osf.io/tz6e7/).

Chapter 9 Exercises

Exercise 9.1: From Another Perspective: Considering Diversity, Justice, and Sustainability Questions Related to National Parks

- Pick your favorite national park.
- Identify the Indigenous people who lived in that area before it was settled into the United States.
- Research the conditions of the descendants of those people today.
- How was the land acquired by the US government?
- Research the ecosystem that is represented in the national park, and review the sustainability of that given any development surrounding the park.

- Consider national park management and how the park could be managed differently to uplift diversity, social justice, and sustainability?

Exercise 9.2: Uplifting Everyone: Considering Book Question and DJS

- The main hypothesis ignores subject characteristics when predicting that age and difficulty would predict miles hiked.
- In what ways would diversity add complexity to the predictions?
  - To what degree is the research question limited by known biases of the researcher?
  - Consider common diversity categories (ethnicity, gender, health status).
    - No estimated effect for ethnicity or gender, but
    - Healthy should outperform unhealthy hikers.
  - Consider uncommon categories (residential scenery, parents' hobbies).
    - Growing up seeing national parks should increase miles hiked.
    - People who were raised by parents who hiked will hike more miles.
- In what ways would social justice add complexity to predictions?
  - Will the project benefit society in an equitable manner?
  - For the project, consider systems at the park (e.g., Is there bias in who can visit?).
  - For the project, consider systems about the person (e.g., Was there bias that benefitted or impaired their hiking ability?).
- In what ways would sustainability add complexity to predictions?
  - Will the project benefit sustainability of the planet or humanity?
  - For the project, consider systems at the park (e.g., terrain and climate conditions).
  - For the project, consider systems of the project (e.g., minimal impact on environment).
  - For the project, consider sustainability of the person (e.g., to what degree is hiking intrinsically rewarding to participants?).

Exercise 9.3: Individual Commitment: Considering Student Project From a Diversity, Social Justice, and Sustainability Lens

Does your research question explicitly test questions related to diversity, social justice, and sustainability? If yes, this might still be illuminating. If no, this might help you consider limitations and future directions.

- In what ways would diversity add complexity to the predictions?
  - To what degree is the research question limited by known biases of the researcher?

- Consider common diversity categories.
- Consider uncommon categories.

- In what ways would social justice add complexity to predictions?
    - Will the project benefit society in an equitable manner?
    - For the project, consider systems at local institution.
    - For the project, consider systems about the person.
- In what ways would sustainability add complexity to predictions?
    - Will the project benefit sustainability of the planet or humanity?
    - For the project, consider systems at local institution.
    - For the project, consider systems of the project.
    - For the project, consider sustainability of the person.
- To the degree that answers and considerations reflected limitations of the current study, add that to the Discussion section, and think about ways these issues could be tested in future research.

END-OF-CHAPTER REVIEW QUESTIONS

1. Who were three of the first African American women to earn psychology Ph.D.s?
2. Define WEIRD.
3. Why are WEIRD populations problematic in research?
4. Describe one way that psychology subjugated a portion of the population.
5. Explain the connection between stereotypes, prejudice, and discrimination.
6. Name three ways that researchers can promote each facet of diversity, justice, and sustainability in their research practices. Are there any contradictions?
7. In what ways can research be sustainable for both the participants and the researchers?

References

Aptheker, H. (1991). Anti-racism in US history: The first two hundred years. *Nature, Society, and Thought*, 4(4), 463.

Auchincloss, L. C., Laursen, S. L., Branchaw, J. L., Eagan, K., Graham, M., Hanauer, D. I., . . . Dolan, E. L. (2014). *Assessment of course-based undergraduate research experiences: A meeting report*. Retrieved from https://www.lifescied.org/doi/full/10.1187/cbe.14-01-0004

Azar, B. (2010, May). Are your findings "WEIRD"? *Monitor on Psychology, 41*(5). Retrieved from www.apa.org/monitor/2010/05/weird

Bayer, R. (1987). *Homosexuality and American psychiatry: The politics of diagnosis*. Princeton: Princeton University Press.

Bertrand, M., & Mullainathan, S. (2004). Are Emily and Greg more employable than Lakisha and Jamal? A field experiment on labor market discrimination. *American Economic Review, 94*(4), 991–1013.

Bhambra, G. K., Gebrial, D., & Nişancıoğlu, K. (2018). *Decolonising the university*. London: Pluto Press.

Blight, D. W. (1999). *David Blight on racism in the abolitionist movement*. Retrieved February 3, 2021, from www.pbs.org/wgbh/aia/part4/4i2978.html

Brigham, C. C., & Yerkes, R. M. (1923). *A study of American intelligence*. Princeton: Princeton University Press.

Carter, R. T., Lau, M. Y., Johnson, V., & Kirkinis, K. (2017). Racial discrimination and health outcomes among racial/ethnic minorities: A meta-analytic review. *Journal of Multicultural Counseling and Development, 45*(4), 232–259.

Cohen-Almagor, R. (2018). Taking North American white supremacist groups seriously: The scope and the challenge of hate speech on the Internet. *International Journal of Crime, Justice, and Social Democracy, 7*(2), 38–57.

Cook, S. W. (1990). Toward a psychology of improving justice: Research on extending the equality principle to victims of social injustice. *Journal of Social Issues, 46*(1), 147–161.

Cramblet Alvarez, L. D., Jones, K. N., Leach, J. L., & Rodriguez, J. L. (2020). Unsung pioneers of psychology: A content analysis of who makes history (and who doesn't). *American Journal of Psychology, 133*(2), 241–262. doi:10.5406/amerjpsyc.133.2.0241

Cynkar, A. (2007). The changing gender composition of psychology. *Monitor on Psychology, 38*(6), 46.

Darmstadt, A., Prinz, M., & Saal, O. (2019). The murder of Keira: Misinformation and hate speech as far-right online strategies. *Journal of Strategic Security, 4*(4), 151–166.

Darwin, F. (1919). A phaenological study. *New Phytologist, 18*(9), 287–298.

Duarte, J. L., Crawford, J. T., Stern, C., Haidt, J., Jussim, L., & Tetlock, P. E. (2015). Political diversity will improve social psychological science. *Behavioral and Brain Sciences, 38*.

Fassinger, R., & Morrow, S. L. (2013). Toward best practices in quantitative, qualitative, and mixed-method research: A social justice perspective. *Journal for Social Action in Counseling & Psychology, 5*(2), 69–83.

Furnham, A., Richards, S. C., & Paulhus, D. L. (2013). The dark triad of personality: A 10 year review. *Social and Personality Psychology Compass, 7*(3), 199–216.

Garth, T. (1934). The problem of race psychology: A general statement. *The Journal of Negro Education, 3*(3), 319–327. doi:10.2307/2292376

Goodenough, F. L. (1926). Racial differences in the intelligence of school children. *Journal of Experimental Psychology, 9*(5), 388–397. doi:10.1037/h0073325

Grahe, J. E., Cuccolo, K., Leighton, D. C., & Cramblet Alvarez, L. D. (2019). Open science promotes diverse, just, and sustainable research and educational outcomes. *Psychology Learning & Teaching, 19*(1), 5–20.

Guthrie, R. (2004). *Even the rat was white: A historical view of psychology*. Boston, MA: Pearson.

Haidt, J., & Jussim, L. (2016). Psychological science and viewpoint diversity. *APS Observer, 29*(2).

Henrich, J., Heine, S. J., & Norenzayan, A. (2010). The weirdest people in the world? *Behavioral and Brain Sciences, 33*(2–3), 61–83.

Herrnstein, R. J., & Murray, C. (1994). *Bell curve: Intelligence and class structure in American life*. New York: Simon and Schuster.

Lerner, M. J. (1977). The justice motive: Some hypotheses as to its origins and forms. *Journal of Personality, 45*(1), 1–52. doi:10.1111/j.1467-6494.1977.tb00591.x

Moshontz, H., Campbell, L., Ebersole, C. R., IJzerman, H., Urry, H. L., Forscher, P. S., . . . Chartier, C. R. (2018). The psychological science accelerator: Advancing psychology through a distributed collaborative network. *Advances in Methods and Practices in Psychological Science, 1*(4), 501–515.

O'Connell, A. N., & Russo, N. F. (Eds.). (1983). *Models of achievement: Reflections of eminent women in psychology*. New York: Columbia University Press.

Petty, R. E., & Cacioppo, J. T. (1984). The effects of involvement on responses to argument quantity and quality: Central and peripheral routes to persuasion. *Journal of Personality and Social Psychology, 46*(1), 69.

Redding, R. E. (2001). Sociopolitical diversity in psychology: The case for pluralism. *American Psychologist, 56*(3), 205.

Rosenthal, R., & Jacobson, L. (1968). Pygmalion in the classroom. *The Urban Review, 3*(1), 16–20.

Saini, A. (2017). *Inferior: How science got women wrong and the new research that's rewriting the story*. Boston: Beacon Press.

Schwartz, S. H. (1992). Universals in the content and structure of values: Theoretical advances and empirical tests in 20 countries. In *Advances in experimental social psychology* (Vol. 25, pp. 1–65). New York: Academic Press.

Schwartz, S. H. (2012). An overview of the Schwartz theory of basic values. *Online Readings in Psychology and Culture, 2*(1), 2307–0919.

Swim, J. K., & Bloodhart, B. (2018). The intergroup foundations of climate change justice. *Group Processes & Intergroup Relations, 21*(3), 472–496.

Syed, M., Santos, C., Yoo, H. C., & Juang, L. P. (2018). Invisibility of racial/ethnic minorities in developmental science: Implications for research and institutional practices. *American Psychologist, 73*(6), 812.

Woodworth, R. S. (1916). Comparative psychology of races. *Psychological Bulletin, 13*(10), 388–397. doi:10.1037/h0075983

10 SCIENTIFIC TRANSPARENCY

A Theme for a Movement

Chapter 10 Objectives

- Define scientific transparency
- Explore the thought experiments of the Scientific Utopia series
- Discuss paths forward and obstacles in the way of open science
  - Opening communications
  - Restructuring incentives
  - Crowdsourcing science
  - Obstacles to open science
- Explore the future of open science
- Present ways to share research beyond the classroom

MUSIC IN A BOX: ABOUT THE SONG "SCIENTIFIC TRANSPARENCY"

"Scientific Transparency" is the final song in the concept album, representing a hopeful ending to the Replication Crisis and the beginning of the next phases of scientific methodology. The song celebrates open science as an inspirational ideological answer that excites others to work together, to openly and respectfully disagree, and to continue to improve their own research practices. The lyrics also recognize the causes for some open science criticism and invite reconciliation. As the lessons come to an end, this song reminds the listener of the basic themes presented throughout the album. "Scientific Transparency" is the true closing song, composed of complex chords and unusual arrangements that signify a resolution and an inclusive, diverse path forward.

SCIENTIFIC TRANSPARENCY AS THE PATH FORWARD

When the US Congress passed and President Ulysses S. Grant signed the Yellowstone National Park Protection Act in 1889 and created Yellowstone National Park, they presented the world with a new model for national land and resource management. Today, the collection of national parks resources showcases the cultural wealth of our country. Spawning a worldwide phenomenon, there are 3,283 national parks in more than 183 countries that followed the US example. Their value is unequivocal; they cannot be replaced. These spaces protect rare ecosystems while providing necessary connections with history and nature to millions of people each year. Yet their beauty also is a facade. Their creation represents a protection of spaces for one group (US citizens) at the cost of the civilizations that lived in those spaces and acted as stewards for millennia before that. The descendants of those people still live with us. Victims of attempted genocide, broken promises, and centuries of misrepresentation in media and history, they still survive. I feel compelled to face this part of our history, especially when we visit these spaces.

Beyond their challenging history, many of these spaces are also under threats from climate change and/or unchecked human development. One example is the repeated wildfires in Glacier National Park. They create heart-wrenching sadness for me, because while forest fires are a natural and healthy part of the rejuvenation of a forest, these are too extreme and too frequent. Another example is Everglades National Park in Florida. Not only is the natural system of wetlands being carved up for urban and agricultural development, but the freshwater necessary to keep the system healthy is reduced while the saltwater creeps further up the shores as sea levels rise. Flora and fauna have adapted and survived global weather changes in the past, but their corridors to escape and regroup are closed by the relentless growth of humans around them. We are compelled to visit these spaces because they are rare and precious, but at the same time our pilgrimages do damage and threaten their health. Our responsibility as citizens is to work together with the government to act as stewards of our natural and cultural resources. This sometimes conflicts with our personal goals as we try to make our individual ways through life. The solutions to the community problem are as multifaceted as solutions to other social justice problems, and sometimes the path seems insurmountable. However, much like a long backcountry hike, we must continue toward a destination.

Again, the national parks provide an ample metaphor for this discussion regarding scientific transparency because the archives of published work represent our collective scientific knowledge. And yet, that compendium of work represents a façade. I am fond of repeating two things I believe about science: (1) science never represents truth, only our best representation of truth as we currently understand it, and (2) science is iterative; each wave of science provides new information which should be closer to a true reality. But we know that our practices led to increased representation of false findings. Because scientific findings always include some degree of ambiguity regarding alternative explanations, we can never know which aspects of our compendium are true or false. Only through rigorous repeated testing can true effects be separated from false effects. Scientific transparency represents the path forward, but it too has drawbacks and challenges. This final chapter offers a final effort to achieve the ideal by first reviewing principles of the Scientific Utopia outlined by Nosek and others (Nosek & Bar-Anan, 2012; Nosek, Spies, & Motyl, 2012; Uhlmann et al., 2019) then considering major obstacles to reaching it and finishing with a look toward an imagined future reality.

VISITING SCIENTIFIC UTOPIA

One of the most impressive characteristics of Brian Nosek is his positive idealism. In his role as a leader of the open science movement, he has been receptive and responsive to criticisms while encouraging those in his sphere to approach the movement. As the conversation regarding problems in psychology research emerged, he and his colleagues offered a path away from the problems. While leveling strong criticisms regarding the publication and promotion processes, they also offered solutions to reach their version of Scientific Utopia in a pair of papers (Scientific Utopia I, Nosek & Bar-Anan, 2012; Scientific Utopia II, Nosek et al., 2012). These papers address topics related to scientific communication and incentives and structures, respectively. Notably, they then went into action trying to achieve that utopia with the facilitation of the open science collaboration, the formation of the Center for Open Science, and the litany of subsequent efforts that were documented earlier in the text. More recently, Scientific Utopia III (Uhlmann et al., 2019) focused on crowdsourcing science, which was one practice introduced in the Scientific Utopia II paper.

This book represents an attempt to explain the present state of open science movements, and since we expect scientific practices to continue to evolve, it should be interesting to visit this idealist Scientific Utopia and consider how the journey might continue toward it. Of course, a diverse population of scientists likely envision a multiverse of Scientific Utopias, but the ideals set forth here has been embraced by large networks of supporters. To maintain the image of the potential utopia, the ideals will be presented independently of any efforts made to achieve it or impediments to its implementation. Once presented, I will consider what steps have been made to achieve the ideals as well as impediments that remain.

BUILDING A SCIENTIFIC UTOPIA

Scientific Utopia I: Opening Communication Systems

Nosek and Bar-Anan (2012) offer six ideals that would represent open scientific communication in a utopian field of research. As a first step, they recommend (1) fully embracing digital communication because it provides broader access with less cost. Noting that publishers still charge for digital modalities, they also recommend (2) open access to all published research. This would increase dissemination of content while simultaneously reducing superfluous research costs from exorbitant subscription fees. When it becomes easy to self-publish, then (3) professional evaluation can be disentangled from publication. This can be achieved by measuring the impact of an article rather than the quality of the outlet in which it was published.

They assert that once these first three ideas are achieved, the next step is to replace the current model with (4) a graded evaluation system of reviews in which publishers promote rather than publish findings. In other words, rather than an accept or reject review, manuscripts receive a rating or grade on some scale. Then editors could select from the best findings and encourage them. Their next steps recognize that the important service peers offer during the review process should also be more transparent. By (5) publishing peer review, they argue that reviewing would be better incentivized for both participation and quality. Finally, they argue that the review process should be ongoing and that systems should (6) allow for ongoing, continuous peer review. Rather than the present process, in which a manuscript attains a level of superior status once published, continuous peer review recognizes that a select few reviewers have a limited perspective. Additionally, this process offers the possibility of specializing expertise such that a class of researchers might focus on reviewing rather than conducting and publishing research. Such a system recognizes and values all components of the research process as equal.

Collectively, these recommendations center on how to change the research once it is completed. Other than expectations that some researchers might focus their time primarily as reviewers if open reviewer systems develop, there is little that would change for how the research itself was conducted. These recommendations would provide for sharing research with increased speed and breadth of dissemination. Instead, recommendations for the conduct of research itself follow in the second utopia installment.

Table 10.1 Collection of Utopian Recommendations

Utopian Recommendation	*Corresponding Initiative*	*Grahe's 25-Year Prediction*
CO1. Fully embracing digital communication	Data repositories (OSF, Dataverse, Figshare)	Fully achieved
CO2. Open access to all published research	Preprint servers, PsyArXiv Open-access policies	More access but not complete
CO3. Professional evaluation can be disentangled from publication	Author contribution lists	Different challenges
CO4. A graded evaluation system of reviews		Some examples but not universal
CO5. Publishing peer review	Publons	Some examples but not universal
CO6. Ongoing, continuous peer review	PsyArXiv	Some examples but not universal
IN1. Promoting and rewarding paradigm-driven research	Reproducibility Project Many Labs projects	Many examples but not universal
IN2. Checklists to encourage transparency	TOP guidelines	Fully achieved
IN3. Interrupting dysfunctional incentives		Many examples but not universal

(*Continued*)

Table 10.1 (Continued)

Utopian Recommendation	*Corresponding Initiative*	*Grahe's 25-Year Prediction*
IN4. Replication metrics to guide study	The Replication Recipe	Fully achieved
IN5. Crowdsourcing replication efforts	A plethora of examples (e.g., RP:P, CREP, PSA)	More diverse examples
IN6. Journals focused on the soundness of research	Journal signatories of TOP guidelines	Many examples but not universal
IN7. Lowering or removing barriers to publication	PsyArXiv	Many examples but not universal
IN8. Open data	Open Data Badges	Many examples but not universal
IN9. Open materials	Open Materials Badges	Many examples but not universal
IN10. Open workflow	Preregistration	Many examples but not universal
CR1. Changing distribution of grant funding		Some examples but not universal
CR2. Adopting author contribution checklists	Adopted by some journals & projects	Many examples but not universal
CR3. Adjusting hiring and promotion criteria		Different challenges
CR4. Integrating crowd science into pedagogy	CREP, Open Stats Lab, open science syllabi	Many examples but not universal
CR5. Changing publication criteria	New journals	Many examples but not universal
CR6. Developing improved infrastructure	OSF, PSA	Fully achieved

Note: CO-communications, IN-incentives, CR-crowd

Scientific Utopia II: Restructuring Incentives and Practices

In this second installment, Nosek et al. (2012) present evidence demonstrating that the traditional research model values novelty and publishability over rigor and reproducibility. Couched in the context of the replication crisis, they then reviewed a series of changes, which they argued would not lead to better science. Many of these are existing practices or ones tried in the past including: conceptual replications, journals devoted to null findings, educational campaigns encouraging replication, increasing reviewer expectations to find errors, and making it more difficult to publish to reduce malfeasance. Collectively, they found the existing outcomes to be too minimal and the likelihood for change nonexistent.

Instead of these failed efforts to catch bad science, they offered 10 strategies they asserted would advance knowledge accumulation: (1) promoting and rewarding paradigm-driven research, (2) checklists to encourage transparency, (3) interrupting dysfunctional incentives, (4) replication metrics to guide study, (5) crowdsourcing replication efforts, (6) journals focused on the soundness of research, (7) lowering or removing barriers to publication, (8) open data, (9) open materials, and (10) open workflow. These recommendations try to ease researchers into a transparent science with rewards. And yet, in contrast to the publishing ideals, most of these include major shifts from traditional models. Taking each in turn, it is easy to see both their reasoning and some initial costs.

Directed at methods themselves, the idea of (1) promoting and rewarding paradigm-driven research would value extending theory by making systematic changes to the methodology to determine the boundary conditions where the effects exist or do not. They assert that directed attempts such as this are better than valuing chaotic tests of theory with disconnected conceptual replications. So that these efforts are directed to the most important research, they recommend that (4) replication metrics are developed to determine which effects need to be replicated. For instance, using published sample sizes and p-values along with impact metrics of the articles, effects that have greater impact but have dubious reproducibility should be studied first. Coupled with (5) crowdsourcing research, scientists can engage in large-scale tests of effects and enable metascience. Simultaneously, crowds are incentivized to offer small contributions to the group in exchange for major impacts of the study.

To encourage greater participation in these crowd projects focused on paradigm-driven research, they recommend (3) interrupting dysfunctional incentives by disconnecting the expectation that a good scientist is one who publishes first-author papers in top-tier journals. They suggest that early career researchers would benefit from a systematic review of how hiring criteria are related outcomes. Related, they suggest that (6) journals focused on the soundness of research rather than the importance of research would train their reviewers to focus on the methods and reproducibility. Further, coupled with the use of repositories, they assert that page limits should no longer apply to research and suggest that editors request more supporting documentation. Further, rather than raising barriers to avoid having bad research published, they recommend (7) lowering or removing barriers to publication. This recommendation is assuming a utopian communication system that employs digital repositories to allow open access of any self-published researcher (see Scientific Utopia I).

To guide this system of self-publishing metascientific, paradigm-driven research, they further recommend (2) checklists to encourage transparency. They suggest different lists for authors, reviewers, and editors that are easy to read and offer constant reminders of review criteria and preferred methods of presentation. The capstone of this utopia is transparency of each study in that they recommend expecting the author to (8) open data, (9) open materials, and (10) open procedures as part of the publication process. As we discussed earlier, the objective is to place these items in a public repository and freeze the page in time so that the researcher cannot return and change it later. This allows other researchers to redo analyses or closely replicate the study to make their own determination of the strength of the reported findings.

Collectively, these 10 recommendations offer both easy and difficult changes to implement. Taken together with those from Science Utopia I, these 16 recommendations offer direct guidance on what changes should be implemented to achieve a more transparent science. Both of these papers were forward looking and offered few existing examples of suggestions because there were none. In contrast, the third utopia paper, which focused on crowdsourcing research, followed numerous years of active attempts to complete this type of work.

Scientific Utopia III: Crowdsourcing Science

Uhlmann et al. (2019) offered a literature review both justifying the benefits of crowdsourcing and reviewing existing projects and how they are being used. Much of this was presented in Chapter 2, but here, I will also share the recommendations for structural changes that would enable them to develop even more fully. These recommendations are proffered to support the ideal presented in Scientific Utopia II that more crowdsourcing is idealistic. They assert that crowdsourcing enables metascience that can advance theory while concurrently assessing the robustness of findings using a more democratic approach. To facilitate increased crowdsourcing, they put forth six recommendations: (1) changing distribution of grant funding, (2) adopting author contribution checklists, (3) adjusting hiring and promotion criteria, (4) integrating crowd science into pedagogy, (5) changing publication criteria, and (6) developing improved infrastructure.

The desire for (1) changing distribution of grant funding includes a desire to see both more value placed on this type of research and to whom the funding is offered. Rather than a bulk going to a few scientists, they recommend more even distribution and for more concrete deliverables. For instance, a funding agency can offer rewards for achieving some project-related task (collecting a sample of data) rather than waiting for proposals for research agendas. As these projects proliferate, (2) author contribution checklists offer a clearer assessment of the various efforts. Rather than an ambiguous list of names in which only the first and last position suggest some contribution status, detailed checklists help authors and publishers award appropriate recognition to a given project. Rather than a list of authored manuscripts, a candidate for a job or promotion could offer a summary of authored checklists. Someone who makes regular contributions to the study design, theory development, and/or conducting analyses would likely be valued more than someone who made a series of minor or supplemental contributions to manuscripts. Naturally, the candidate needs a conducive audience. The degree to which these contributions are valued might require (3) adjusting hiring and promotion criteria. Individuals in a "publish or perish" work environment (early-career faculty who may work in programs that do not value transparency) might find it more difficult to embrace this work compared to people working with relative freedom (people who are fully promoted and/or working in supportive programs).

To develop researchers who can contribute to crowdsourcing, they recommend (4) integrating crowd science into pedagogy. This benefits students, instructors, and science in different ways but requires dedication and resources. To publish the resulting research from these projects, it would be beneficial to engage in (5) changing publication standards. Specifically, their call echoes that from Scientific Utopia to emphasize rigor or novelty, which would then place greater value on the metascience benefits of the crowd project. Finally, continually (6) developing infrastructure to support these projects would enable more and better projects in the future. While early projects have been successful with existing resources, better project management tools can improve quality, and formalized systems can increase democratization of contributions and decisions. Together, these implemented strategies are intended to facilitate more and better science. However, utopias are hard, if not impossible, to achieve in any context. And so these goals for an ideal future need to be contrasted against the obstacles and impediments to their implementation.

OBSTACLES TO OPEN SCIENCE INITIATIVES

All three papers are forthright regarding challenges to implementing these utopian notions. However, the open science movement has advanced many of these idealistic standards in the past decade. This book reviewed many of these in the previous chapters. Here, they are located in the corresponding initiative column of Table 10.1. It is striking that at least one example was available for all but four of the utopia recommendations (CO4: Graded Evaluations of Reviews; IN3: Interrupting Dysfunctional Incentives; CR1: Changing Distribution of Grant Funding; CR3: Adjusting Hiring and Promotion Criteria). For most of these, change cannot be centralized. Even though the Publons program exists now as a way to publish reviews, publishers and editors would need to coordinate in order to adopt graded evaluations of reviews. Adjusting hiring and promotion criteria requires a faculty committee at more than 4,000 institutions across the US to review and replace their current procedures. Similarly, until there is a massive change in how these occur, it will not be possible to successfully interrupt dysfunctional incentives. Finally, though changing how grants are distributed has not occurred, this could occur without seeking change across the system, rather through a single funder that is willing to try something new.

Much like the other categories in which an example is added as a corollary incentive, adding a single grant funder who pilot tested these recommendations would not demonstrate achievement of the utopian recommendation. Instead, it would demonstrate progress toward it. In fact, none of these utopian recommendations are met at an ideal level. Rather, there are efforts to achieve them with work from individuals or organizations. This still represents success on many levels. Considering the brief time from recommendation until initiatives were implemented, the movement has definitely been busy. Of course, my personal bias also draws me to positive conclusions because I personally have tried to advance more than one of these. So while I have a few concerns to raise, we should trust more objective assessments of the capacity for the field to achieve these ideals.

Such critiques are available (Hesse, 2018; Coyne, 2016; Rubin, 2020), and the categories are listed in Table 10.1 to show which initiative they most negatively threaten. Hesse's (2018) critique primarily identified impediments to sharing open data with general categories such as: practicality, lack of cohesive lexicon or ontology, intransigence to tradition, and the dark side of open science. Coyne (2016) criticized replication initiatives specifically where he specified multiple concerns, notably that replication initiatives do not work for some research topics, are not designed to detect fraud, and include numerous unnecessary and unhelpful elements. Rubin (2020) targeted preregistrations as not necessary, arguing that it offers historical transparency, which is not better than contemporary transparency. In fact, Rubin strongly argued in favor of open science generally but questioned the assertions of the effectiveness of preregistration to achieve stated objectives. All three critics actually supported scientific transparency principles but questioned the capacity for the field to achieve the stated goals.

Though Hesse (2018) focused on data, some of the concerns raised could also interfere with broader initiatives. When raising alarms about the practicality of open data (which, in turn, applies to materials and workflow), Hesse lists problems with the accessibility of files such as them becoming obsolete, on being restricted platforms, or being coded inadequately. Additionally there are ethical concerns, because data need to be fully anonymized and IRB permissions for public data can be difficult to obtain, particularly after the fact. Beyond data, concerns raised about a common set of terms or "ontology" could increase difficulty implementing common systems for open or graded reviews because it is difficult to work across subdisciplines. The other concerns reflect more ambiguous challenges. The label "intransigence of tradition" effectively conveys a cascade of obstacles affecting institutions' hiring and promotion practices, journals' policies related to valuing broadly collaborative work, and funding agencies' lack of responsiveness to this new environment (something echoed by Uhlmann et al., 2019). Perhaps these are all subservient to the concerns of individual researchers who feel that they might be scooped if they are open (see also Van Noorden, 2014).

Additionally, Hesse warns about the dark side of open science. He noted the proliferation of new journals with questionable reputations and real concerns about data breaches and the long-term costs of computer storage. To these two "dark side" concerns, I would like to add two more: the politicization of open science in academia and government areas and the bullying against peers that sometimes occurs. The politicization of open science occurred during the Trump presidency when they retroactively applied open science expectations for climate data by refusing to consider any research that did not share data. In this case, the goals of open science generally were used to limit specific arguments that could be made about human-influenced climate change. While I support retroactively making data open, this is not always possible, so we should demand openness in the future but expect little change from the past.

As much as we should avoid letting political principles interfere with science, we should consider the human impact of our science on others. Some of the early conflicts regarding open science were voiced by people feeling that they were victims of professional bullying from people who seemed generally shocked at others' behavior (Schnall, 2014; Cuddy, Wilmuth, & Carney, 2012). Regardless of the reproducibility of an effect, the people who originally documented the effect should receive basic human kindness until they show that they offer none in return. Of greater concern is whether bullying tends to occur more toward women than men. Gruber et al. (2020) noted the need for increased respect for women researchers in the future, as they are more likely to be the victims of sexual harassment and bullying in the workplace. Open science advocates should be aware of these ambiguous and interpersonal obstacles so that they are fairly representing the costs and benefits of adoption. Additionally, advocates should be aware of these concerns so that they do not fall into the trap of accidentally becoming bullies themselves.

PREDICTING THE FUTURE

Nosek and colleagues painted a picture of a Scientific Utopia in which all levels of the research process are transparent and equitable. There are many thousands of researchers who have embraced some facets of this vision by participating in crowd projects or adopting open practices into their own workflows. Table 10.1 shows that this army of open scientists is active in making change. However, there are also many gaps between the current state of affairs and the utopian ideals for transparent science. Science disciplines will continue to refine their methods across their iterative efforts to uncover truth, and it is unlikely that the utopia will emerge as recommended. Because an inherent human characteristic is to be curious about the future, I hope the reader is drawn to make changes and to wonder about what those changes might become.

In Table 10.1, I added some predictions about what I expect in the next 25 years. By doing this task yourself, you think more about what the ultimate goals are and what is required to achieve them. I do not claim clairvoyance, as my prediction to a friend in 1992 that "this electronic mail thing will never catch on" demonstrates. However, with a bit more age and wisdom regarding this topic, I offered the following possible futures (Fully achieved; More access but not complete; Different challenges; Some examples but not complete; Many examples but not complete). Rather than justify individual predictions, I will explain the categories. Readers might identify their own categories or disagree with mine.

The categories I chose mostly reflect the degree to which I expect the field to adopt each recommendation. For instance, I think in 25 years, science will fully adopt digital communication. However, I think there will be more access but still some, if not many, limits to scientific knowledge. For many of these initiatives, entrenchment in traditions can lead some portion of the community to maintain contrary systems. I struggle with expecting that large private companies (social media companies, pharmacies) will ever freely give up their proprietary data and yet they conduct science regularly to advance their profits. In this same perspective of hold-outs, I noted "some examples" or "many examples" for initiatives that will continue to proliferate and augment but will not completely replace existing structures.

Finally, for a couple recommendations, I see continued challenges that will likely manifest in unknown ways. For instance, to the degree that hiring and promotion criteria change for many institutions, it is unlikely that the various private entities will all adopt the same practices. Put in another way, while I hope there is continued progress toward social equity and a sustainable future, I do not think this utopia is a mere 25 years away. As such, I anticipate ongoing if unpredictable challenges in incentive structures. More generally, I expect a bifurcation of scientific research into more open and more closed systems. I predict that the open systems will be more diverse, socially just, and sustainable compared to the closed systems. And I predict a continued proliferation and refinement of crowd projects, but they cannot ever fully replace the science of discovery, which will become smaller in scale.

FINAL CONSIDERATIONS ABOUT THE FUTURE

This book closes by restating four major lessons regarding the status of science for those considering change. These are not particularly related to open science but rather to more basic characteristics of change within the human system. The book is framed around a set of songs that were written to encourage a distinct reflection on how to inspire scientific transparency. In the spirit of that approach, these final lessons are stated here as though they too were lyrics in a song.

Four Major Lessons

There are no permanent solutions.

> Science is only approximate,
>
> but it is better than alternatives.
>
> Humans are all flawed, but we just keep pushing.

Perhaps the reader has a melody in mind for how this stanza would go. I can think of a few melodies that might be fun. But more importantly, these lessons should keep us somewhat humble. The first line states that "there are no permanent solutions" because problems evolve as solutions emerge. For any changes made, there are likely some unintended consequences. Making peer review public is intended to help fix publication bias, but what byproducts of that change will come? Throughout this book, I've repeated the truism that "science is only approximate" and never represents Truth. It makes scientific findings vulnerable to naysayers protecting their profit. Tobacco and oil companies have regularly challenged science findings by overstating this limitation. While scientific arguments are vulnerable to attack, empirical evidence is still stronger than alternative methods to explain human behavior. And that challenge is significant, since "Humans are all flawed, but we just keep pushing." In other words, the scientist asking the questions will lead to inaccuracy if systems do not check likely human errors. As a human yourself, I strongly urge you to embrace such systems, because you are going to make errors.

SHARING RESEARCH BEYOND THE CLASSROOM

For students in a class, I hope that reading this chapter coincides with the final efforts to clean up the writing of a manuscript and/or prepare a presentation using poster or oral formats. If the research project was also intended to be shared beyond the classroom at a conference or as a manuscript submission, that work will likely occur after the course is completed. Given that this book is intended to increase the open science contributions of undergraduates, this final section offers advice on this process. For the final writing of the paper, readers should review the writing advice offered in Chapter 7. Here, the goal is to prepare the project for other outlets. Table 10.2 shows a list of possible outcomes, the relative challenge to get it accepted, and the minimum needed to share the finding.

Poster and Oral Preparation

Posters are primarily visual presentations of the research project. While my first posters consisted of printing sections on white paper and gluing them to colored backing, today, most posters are constructed using presentation software, such as Microsoft PowerPoint or Google Slides. By setting the size of the slide to a larger size (2′ × 3′, 3′ × 4′, or 4′ × 6′ most likely), authors populate content with fonts large enough to see from a few feet away and with visual depictions of the design and findings wherever possible. As technology makes it easier to connect researchers, authors no longer carry "reprints" of their posters to hand out at the conference. Rather, they collect email addresses or, more recently, add QR codes to offer increased access to the content there. Embracing this technology and encouraging greater interactions between the researcher and the attendee, Pedwell, Hardy, and Rowland (2017) offered guidance on building posters that followed better design features visually.

Table 10.2 Where to Share Research

Public repository	No review	Data, material, summary
Student or regional conference	Low bar	Abstract, poster, or talk
National, international	Medium bar	Abstract, poster, or talk
Student journal	Medium bar	Complete manuscript
Low-ranking journal	High bar	Complete manuscript
High-ranking journal	Very high bar	Complete manuscript

These posters use even fewer words and rely more on visuals, but they also offer electronic connections to repositories of information and more detailed explanations.

Preparing slides for oral presentation or talks requires more slides and some coordination with the spoken presentation. In this case, authors should be careful to plan their talk so they do not have too much content, and they should always have some slides they are willing to skip should they need to cut their talk short in the middle of presenting. As with posters, there are recommended design standards to make work accessible to diverse audiences. For example, slides should avoid overuse of color to avoid confusing color-blind audience members. Presenters will also find the audience to be more alert and receptive if the content of the talk augments rather than repeats the slides. Reading content directly from a slide offers little added value. This can be achieved by using images that remind the speaker of content, but allow the audience to visualize, rather than read, the content. Finally, I recommend about one slide per minute on average. If your talk exceeds that limit, practice again to verify timing is correct.

Presenting Posters and Talks

If I had any natural skill at guitar, or at least an earlier start, I would have pursued a rock-n-roll future more fervently. This is relevant because I LOVE public speaking. It's a real thrill to share my excitement about some topic with an audience. I recognize that this is a minority perspective, and I empathize with speakers who are naturally introverted or socially anxious. To those readers, I offer these thoughts and recommendations to help make the experience less dreadful and maybe even exciting.

There are ways to control the dread, like by taking advantage of misattribution of physiological arousal. The first stage of the two-factor theory of emotion is arousal in response to a stimulus; the second stage is labeling that arousal as an emotion (Schachter, 1959). For those who label the arousal of public speaking as exciting, the event is more likely going to be fun (Brooks, 2014). For those who label it as fear, the event will become anxiety provoking and threatening. It might not happen the first time, but with practice, public speaking can be fun.

Another way to reduce dread is to practice, practice, practice. Any behavior that is well learned is more likely to be expressed successfully during a stressful or ambiguous situation (Ayres, Schliesman, & Sonandré, 1998). This principle is called social facilitation. Conversely, if a talk is not well learned, the presence of others can lead to poor performance or social inhibition (Bond & Titus, 1983). Researchers who spent effort reading and thinking about a project know the literature and justification better than most audience members. And researchers who carefully tested hypotheses and examined their analysis protocols and data closely can answer questions about the outcomes. Yet the best talks should still elicit challenging questions. If there is no debate or controversy at all, what is the point of the question? This does not mean that the discussion needs to be contentious, but researchers should prepare talks by trying to anticipate likely questions and even preparing answers for them.

However, researchers who know their material should feel confident that they can address any question that arrives. But sometimes, not only is a question unexpected, but the researcher is also unprepared to answer on the spot. Two points about difficult or impossible questions. First, unless the question is approached in a hostile way, thank the audience member who asked it. Good questions advance science, particularly if it is a question that was not previously anticipated.

Second, if the reader promises to keep a secret, I will share three answers that, when used correctly, can get any speaker out of any impossible-to-answer question. Now, these are powerful tools and, if used incorrectly, will cause awkward situations. The trick is to carefully listen to any question, and then answer it to the best of your ability. If there is no answer, use one of these: (1) That is an interesting question, but it is beyond the scope of the present evidence; (2) That is something I have not yet answered myself, but I will look into it; or (3) That is a great point, what do you think? Note that in each case, the speaker is not trying to pretend to be more knowledgeable than they are, but rather, they are trying to make the most of the situation by either keeping the topic focused on the research at hand or by extending the conversation until more is learned.

Converting Coursework to Manuscript Submissions

For researchers intending to submit their class project to a journal, even a student journal, there is much work to be done after the course is complete. The level of rigor expected of a student paper is vastly different than the rigor expected of published work. Before submitting a paper, please follow the, "edit, edit, and edit again" mantra. But also spend some time identifying the right outlet for the work being considered. All journals list their aims and scope, which specifies the types of papers they seek. Further, they give detailed instructions about the paper format. It is a good idea to read a few recent papers from the target journal to become familiar with expectations. Follow the guidance of the course instructor or other mentors to be sure that the journal is a good outlet for the research. Finally, be patient with a process that can take months or even years.

With Employers and Yourself

The vast majority of psychology students are not going to publish their work, even as a conference presentation. Rather, most psychology majors go into fields as research consumers rather than research generators, as discussed in Chapter 7. For those students, the final project offers evidence of excellence and passion. Even if the project is not directly related to an intended career, it might be that showing a well-organized, professional product can display skills desired by the employer. This is not always going to be the case, but when it is, the open science procedures offer a ready-made project to share with the employer.

But the one other person who might want to reflect on this work in the future is you. After 25 years, I enjoyed reading work I did as a student, even if the work is below my standards of today. The educational process feels long to young students, but it is a brief episode in our lives. These projects will be present and offer ready-made memories to help us reflect on the events that helped us learn to become who we eventually became. Much like the predictions of the larger process of open science, I expect that after 25 years, reflecting on this work will mostly bring memories of joy and pride but also reignite the trials and challenges that make this learning experience so impactful.

Chapter 10: Crisis Schmeisis Journey Book Review

The Captain's Dog

Roland Smith

This children's book takes the perspective of a dog that was on the journey with Lewis and Clark when they traveled west. The narrative is a bit optimistic, but it still reflects the serious nature of this early act of settling the west. This chapter also offers an optimistic but serious view of the future (https://osf.io/tz6e7/).

CHAPTER 10 EXERCISES

Exercise 10.1: Difficult Transparency Conversations: Practice Identifying the Critiques and Responses to Common Open Science Criticisms

- With a partner, randomly assign one partner to play the role of open science advocate and the other to play an open science skeptic.
- Take turns discussing both positions for each of the following conversation prompts.
    - Transparency increases the possibility of being scooped.
    - Replications cannot definitively tests original effects.
    - Sharing data and materials makes it possible for bullies to target or harass.

Exercise 10.2: Complete a Transparency Report

- How to use the Transparency Checklist from Aczel et al. (2020)
    - www.shinyapps.org/apps/TransparencyChecklist/
    - Aczel et al. (2020) www.nature.com/articles/s41562-019-0772-6
- These are the instructions provided by the creator.
- The checklist refers to a single study of interest.

- Please respond to each checklist item. If necessary, you can provide an explanation at the end of each section.
- When the question refers to your manuscript, this includes all supplementary materials and appendices that are relevant to the study of interest.
- After all the questions have been answered, you can generate a transparency report for your study by pressing the button labeled GENERATE REPORT at the bottom of the page.
- Save your transparency report on your computer. Note that after you download your report, your responses on the checklist will not be saved by our webpage.
- Upload your transparency report to a public repository.

Exercise 10.3: Finding an Outlet: Identify Ideal and Practical Publication Outlets

- Identifying the best outlet is most of the challenge in sharing our research. Follow the decision tree to see the best fit.
- Is there IRB approval to publish data?
  - If no, then stop.
- Is research part of a crowd project?
  - If yes, then check with project instructions regarding sharing.
- Is there time to present at a regional, national, international conference?
  - If yes, these are great opportunities to share research with an audience, and it does not limit future paper publication.
- How strong is the theoretical contribution? Is the research advancing novel theory or testing established theory?
  - If no, check journal instructions for good fit.
- Identify journals that match content.
  - Do a Google search in the specialty subfield and review lists of journals
    - "list of psychology journals social," "list of psychology journals cognitive"
- How good are the methods?
  - Is research worthy of publication by demonstrating an effect with high power, strong reliability, and convincing validity?
  - If no, then stop. Don't submit research that cannot draw strong inferences.
- If study has good methods with IRB approval, check out Taylor and Francis Journal Finder:

- https://authorservices.taylorandfrancis.com/publishing-your-research/choosing-a-journal/journal-suggester/?utm_medium=email&utm_source=EmailStudio&utm_campaign=JPA18388+-+week+1_3906527 Submit abstract to journal submission portal.

END-OF-CHAPTER REVIEW QUESTIONS

1. List the six ideals described in Scientific Utopia I.
2. What are five of the strategies offered by Scientific Utopia II to advance knowledge accumulations?
3. List three of the recommendations from Scientific Utopia III to facilitate increased crowdsourcing.
4. List the ways to share research beyond the classroom assignment.
5. What is one way to get better at public speaking? Do you feel confident or anxious about speaking in public?
6. What is one piece of advice that will help you the most in regard to sharing your research publicly?
7. Which park was the first national park in the United States?
8. How is the creation of the National Park Service similar to the creation of the open science movement?
9. What are some ways the National Park Service supports diversity, justice, and sustainability that are also foundational to the open science movement?
10. Name three utopian recommendations and the corresponding initiative(s) that are seeking to achieve those goals.
11. Name two initiatives that aim to achieve multiple utopian goals.

References

Aczel, B., Szaszi, B., Sarafoglou, A., Kekecs, Z., Kucharský, Š., Benjamin, D., . . . Wagenmakers, E. J. (2020). A consensus-based transparency checklist. *Nature Human Behaviour*, *4*(1), 4–6.

Ayres, J., Schliesman, T., & Sonandré, D. A. (1998). Practice makes perfect but does it help reduce communication apprehension? *Communication Research Reports*, *15*(2), 170–179.

Bond, C. F., & Titus, L. J. (1983). Social facilitation: A meta-analysis of 241 studies. *Psychological Bulletin*, *94*(2), 265.

Brooks, A. W. (2014). Get excited: Reappraising pre-performance anxiety as excitement. *Journal of Experimental Psychology: General*, *143*(3), 1144.

Coyne, J. C. (2016). Replication initiatives will not salvage the trustworthiness of psychology. *BMC Psychology*, *4*(1), 1–11.

Cuddy, A. J., Wilmuth, C. A., & Carney, D. R. (2012). *The benefit of power posing before a high-stakes social evaluation*. Harvard Business School Working Paper Series# 13-027. Boston, MA: Harvard Business School.

Gruber, J., Mendle, J., Lindquist, K. A., Schmader, T., Clark, L. A., Bliss-Moreau, E., . . . Williams, L. A. (2020). The future of women in psychological science. *Perspectives on Psychological Science*, September. doi: 10.1177/1745691620952789

Hesse, B. W. (2018). Can psychology walk the walk of open science? *The American Psychologist*, *73*(2), 126–137. doi:10.1037/amp0000197

Nosek, B. A., & Bar-Anan, Y. (2012). Scientific Utopia: I. Opening scientific communication. *Psychological Inquiry*, *23*(3), 217–243.

Nosek, B. A., Spies, J. R., & Motyl, M. (2012). Scientific Utopia: II. Restructuring incentives and practices to promote truth over publishability. *Perspectives on Psychological Science*, *7*(6), 615–631.

Pedwell, R. K., Hardy, J. A., & Rowland, S. L. (2017). Effective visual design and communication practices for research posters: Exemplars based on the theory and practice of multimedia learning and rhetoric. *Biochemistry and Molecular Biology Education*, *45*(3), 249–261.

Rubin, M. (2020). "Repeated sampling from the same population?" A critique of Neyman and Pearson's responses to Fisher. *European Journal for Philosophy of Science*, *10*(3), 1–15.

Schachter, S. (1959). *The psychology of affiliation: Experimental studies of the sources of gregariousness*. Stanford: Stanford University Press.

Schnall, S. (2014). Clean data: Statistical artifacts wash out replication efforts. *Social Psychology*, *45*(4), 315–317.

Uhlmann, E. L., Ebersole, C. R., Chartier, C. R., Errington, T. M., Kidwell, M. C., Lai, C. K., . . . Nosek, B. A. (2019). Scientific Utopia III: Crowdsourcing science. *Perspectives on Psychological Science*, *14*(5), 711–733.

Van Noorden, R. (2014). Confusion over publisher's pioneering open-data rules. *Nature News*, *515*(7528), 478.

INDEX

Page numbers in *italic* indicate a figure and page numbers in **bold** indicate a table on the corresponding page.

AAC&U *see* Association of American Colleges and Universities (AAC&U)
academic and research publishing, freedom of press in 86–88
academic integrity 84, 90
Aczel, B. 162
American Psychological Association (APA): ethical guidelines 88–92; general principles 89; recommendations on writing 90–91; research and publication 89–90
ANOVA, frequentist statistics through 53–55
APA *see* American Psychological Association (APA)
Association for Psychological Sciences 112
Association of American Colleges and Universities (AAC&U) 100
Auchincloss, L. C. 103, 142

Bangera, G. 101
Bar-Anan, Y. 150
Basic Value Survey 102
Bayesian statistics 55–56
Bem, D. J. 3, 32
Berkeley Institute for Transparency in Social Sciences (BITSS) 111, 112
bias: confirmation 67; experimenter 72; implicit 137–138; in-group 138; publication 2, 77, 120, 142; sample 52–53
biology 137–138
BITSS *see* Berkeley Institute for Transparency in Social Sciences (BITSS)
Blouw, P. 35
Brady, W. J. 114–115
Brownell, S. E. 101
Buckwalter, W. 35

careers: with National Park Service 100; and research projects 96–97
Center for Open Science (COS) 9, 32, 110, 112, 114, 116, 150
Central Limit Theorem 16
Chase, C. 40
Christidis, P 99
CI *see* confidence interval (CI)
Clark, M. P. 139
cluster sampling 50–51, *51*
Collaborative Replications and Education Project (CREP) 20, *20*, 23–25, 27–28, 34, 35, 97, 104, 117–118, 122
Collabra journal 112
Committee on Publication Ethics (COPE) Publication Guidelines 86–88, **87**
confidence interval (CI) 56–59, *57*, **58**
confirmation bias 67
confirmatory and exploratory hypotheses 33–34
Conroy, J. 99
COPE Publication Guidelines *see* Committee on Publication Ethics (COPE) Publication Guidelines
COS *see* Center for Open Science (COS)
course-based undergraduate research experiences (CUREs) 103–105
coursework to manuscript submissions, converting 161
Coyne, J. C. 156
Cramblet Alvarez, L. D. 6, 142
CREP *see* Collaborative Replications and Education Project (CREP)
Crisis Schmeisis 6–7, 28, 44, 60, 77–78, 92–93, 105, 123, 129, 143, 162
Crisis Schmeisis Tour to Increase Scientific Transparency 4
crowdsourcing 26, 39, 113, 154–155
Cuccolo, K. 6, 142
curriculum with CUREs, updating 103

data analysis **87**, 123
decision heavyweights 47–63; Bayesian statistics 55–56; decision-making about 47–48, 59–60; estimation approaches with effect sizes and confidence intervals 56–59, *57*, **58**; frequentist

statistics through ANOVA 53–55; sample bias with random assignment, resolving 52–53; statistical significance of 48–49; testing with unbiased samples 49–52
DEI (diversity, equity, and inclusion) 141
diversity: collaborative replication and 25–28; justice/sustainability and 5, 131, 141–142; open science and 142–143; preregistration and 38–40; race and 133–135; in US National Park System 5–6

EAMMI study *see* Emerging Adulthood Measuring Multiple Institutions (EAMMI) study
EAMMi2 *see* Emerging Adulthood Measuring Multiple Institutions 2: The Next Generation (EAMMi2)
Ebersole, C. R. 115
editing: online 122; the wiki 11; and writing 91
editors: duties of **87**; and reviewers 16, 34, 38, 43, 86, 88, 132
effect sizes/confidence intervals 56–59, *57*, **58**
Emerging Adulthood (2016) 74, 104
Emerging Adulthood Measuring Multiple Institutions (EAMMI) study 74, 118
Emerging Adulthood Measuring Multiple Institutions 2: The Next Generation (EAMMi2) 15, 24, 26, 74, 97, 102, 104, 118, 120, 122
Epskamp, S. 75
estimation approaches, with effect sizes and confidence intervals 56–59, *57*, **58**
ethics/ethical writing 81–83; APA ethical guidelines for 90–91; Grahe's general advice for research reports 91–92; and open science 84–86; and professionalism 83–88; for psychology researchers 88–92
experimenter bias 72
exploratory: and confirmatory hypotheses 33–34; research questions 42–43

Filedrawer Project 114
First Amendment of the US Constitution 86
Flake, J. K. 41, 42
Fleischmann, M. 99
Fosse, N. E. 15
freedom of academic press 86–88
free learning with open access 121–122
frequentist statistics, through ANOVA 53–55
Fried, E. I. 41, 42
front-loading work 36

Gernsbacher, M. A. 110
Google tools, communicating online freely with 122
Grahe, J. E. 6, 15, 25, 91–92, 104, 142
Grant, U. S. 148
Gruber, J. 157

hackathons 112–113
Haidt, J. 132
Haraguchi, A. 132
HARKing 3, 16, 33, 75–76, 82
Harrold, R. 99
Hesse, B. W. 156
history, remembering 65–66
Howard, R. 139
hypotheses 38; confirmatory/exploratory 33–34; generation 10; null 47, 49, 54, **54**, 55, 60; organizing sections according to 92

implicit bias 137–138
in-group bias 138
in-principle acceptance 34
Ioannidis, J. P. 4, 15

JASP *see* Jeffrey's Amazing Statistics Program (JASP)
Jeffrey's Amazing Statistics Program (JASP) 55–56, 123
job market, status of 97–101; careers with National Park Service 100; degrees 98–100, **99**; research experiences 100–101
John, L. K. 82
Journal of Personality and Social Psychology 3
Journal of Social Psychology, The 86
Jussim, L. 132
justice *see* social justice

Kaepernick, C. 141
Keiner, M. 77
Kerr, N. L. 75

Landrum, R. E. 99
Law of Large Numbers 16
Leighton, D. C. 6, 142
Lin, L. 99
LIONESS 122
Loewenstein, G. 82

management theory 115
manuscript submissions 161
Many Children and Many Adolescents projects 116
Many Labs 24, 25, 104, 115–120, 122
Matlab 123
Mende-Siedlecki, P. 114–115
meta-analysis 59, 77
multiverse of p-values 120

National Lampoon 40
Nelson, L. D. 49, 66
Network for International Collaborative Exchange (NICE) 24, 25, 34, 39, 97, 118
NICE *see* Network for International Collaborative Exchange (NICE)
nonrandom sampling 51–52
Nosek, B. A. 70, 112, 115, 149, 150, 152
NPS *see* US National Park System (NPS)
Nuijten, M. 75
Null Hypothesis Significance Testing (NHST) 54

open access, free learning with 121–122
open science: definition of 7–8; and diversity 142–143; and ethics 84–86; initiatives 8–9; obstacles to initiatives 155–157; and social justice 142–143; and sustainability 142–143
open science alphabet 109–125; Collaborative Replications and Education Project 117–118; data analysis 123; early entries, comparing 114–115; inviting a crowd 113–121; National Park System and 113; Network for International Collaborative Exchange 118; Open Sesame 122; past projects, possible uses of 119–121; Psychological Science Accelerator 118–119; resources, finding 111–113; students, as crowd contributors 117; tools 121–123; wave of free online study software 122–123
Open Science Badges 112
Open Science DB 113
Open Science Framework (OSF) 142; Classroom Research Methods Project Template 8–9, *8*; hosting projects on 23; managing projects freely with 122; page 59; software 112; wiki 33
Open Sesame 122
Open Stats Lab 121
oral preparation 159–160
OSF *see* Open Science Framework (OSF)

past research, criticizing 66–68
peer review **87**, 88, 132, 151
Perspective on Psychological Science 2, 114
p-hacking 3, 16, 65–79; future, prediction of 65–66; history, remembering 65–66; questionable research, detecting 76–77; past research, criticizing 66–68
pipeline projects 116
posters 159–161
Prelec, D. 82
preregistration 32–45; benefits to researchers 35–38; choices in OSF *35*; confirmatory and exploratory hypotheses 33–34; confirmatory research questions 42; decision-making 37; definition of 32–34; diversity and 38–40; exploratory research questions 42–43; front-loading work 36; priori capacity, evidence of 37–38; self-deception, reducing 36–37; social justice and 38–40; sustainability and 38–40; timing of 41–43; types of 34–35; undergraduate research project and 43–44; vacation plans in national parks 40–41
press, freedom of 86–88
professionalism, ethics and 83–88
progress, sustaining 138–139
Prosser, I. B. 140
PSA *see* Psychological Science Accelerator (PSA)
PsyArXiv 142
Psychological Science Accelerator (PSA) 24, 25, 35, 39, 118–119, 142–143
Psychological Science journal 35
Psycstudio 122
publication bias 2, 77, 120, 142
"publish or perish" phenomenon 97, 154
p-values 49

Qualtrics 122
questionable research, detecting 76–77

race, and diversity 133–135
random assignment 52–53
Registered Replication Reports (RRRs) 34, 39, 116, 117, 119, 120
Registered Report (RR) 34, 117
Reifman, A. 74
Reinero, D. A. 114–115
Renkewitz, F. 77
replication 14–29, 116; benefits to science 15; contributor management through time and space 23; crisis 1–12, 15; definition of 14–15; diversity 25–28; large-scale 23; National Park System 20–22; problems from single-study projects 16; "right" project, finding 24; social justice 25–28; successful 19–20; sustainability 25–28; types of 16–17, *18*
Replication Crisis 142; definition of 1–2; potential causes of 2–4; reasons for 4
"Replication Recipe, The" (2014) 17, **19**
Reproducibility Project: Psychology (RP:P) 110, 114–116, 120
Reproducibility Projects 112
research: in action 9, 48–49, 81–83, 113; careers 96–97; confirmatory/exploratory questions 42–43; criticizing past 66–68; experiences 100–101; freedom of press in 86–88; and publication 89–90; questionable 76–77; sharing, beyond the classroom 159–162, **159**; social justice in 135–137; sustainability in 139–142; traditional methods courses 102–103; undergraduate projects 43–44; writing advice for 91–92
researcher degrees of freedom, decisions associated with/in 33, 68–76; analysis plan 72–73; conditions and measures 70–71; data collection 71–72; discussion section 75–76; good and questionable science practices **76**; introduction 69–70; methods 70; results 73–74
Research Transparency and Reproducibility Training (RT2) 111
right project, finding 24
Robertson, J. 105
Rosenthal, R. 72, 114
RP:P *see* Reproducibility Project: Psychology (RP:P)
RR *see* Registered Report (RR)
RRRs *see* Registered Replication Reports (RRRs)
RT2 *see* Research Transparency and Reproducibility Training (RT2)
R tool 123
Rubin, M. 156

SAS tool 123
Schwartz, S. H. 134
science 5; crowdsourcing 154–155; replication benefits 15; *see also* open science
scientific method **7**
Scientific Revolution 2.0 15, 32, 66
scientific transparency 7, 148–164; crowdsourcing science 154–155; future prediction and considerations 157–159; incentives and practices, restructuring 152–154; obstacles to open science initiatives 155–157; opening communication systems 150–151; as path forward 148–149; research, sharing 159–162, **159**; utopian recommendations **151–152**; utopia, visiting 150
SEPA *see* South Eastern Psychological Association (SEPA)
Simmons, J. P. 49, 66
Simonsohn, U. 49, 66, 76
single-blinded design 72
SIPS *see* Society for the Improvement of Psychological Science (SIPS)
social facilitation 160
social justice 5; collaborative replication and 25–28; definition of 135; diversity and 141–142; open science and 142–143; preregistration and 38–40; in research context 135–137; in US National Park System 5–6
Society for Personality and Social Psychology 112
Society for the Improvement of Psychological Science (SIPS) 112, 143
Sound of Music, The (1965) 109
South Eastern Psychological Association (SEPA) 102
SPSS tool 123
Stapel, D. 2
Statcheck program 75

stratified sampling 50
Stroebe, W. 116
students, as crowd contributors 117
study design **87**, 154
Study Swap 39
sustainability 5; collaborative replication and 25–28; open science and 142–143; preregistration and 38–40; in research context 139–142; in US National Park System 5–6

talks, presenting 160–161
TOP Guidelines *see* Transparency and Openness Promotion (TOP) Guidelines
Toyokawa, T. 15
traditional research methods courses 102–103
Transparency and Openness Promotion (TOP) Guidelines 7, 84–86, **85**, 88, 112
Turner, A. B. 139
Turri, J. 35

Uhlmann, E. L. 113, 114, 154
undergraduate research, changing 101–105; curriculum with CUREs, updating 103; traditional research methods courses 102–103
US National Park System (NPS) 9; diversity in 5–6; as metaphor for replication 20–22; and open science 113; social justice in 5–6; sustainability in 5–6

Vacation, National Lampoon's 40
Van Bavel, J. J. 114–115
Vazire, S. 112

WEIRD (Western, educated, industrialized, rich, and democratic) 25, 38, 132–134, 139, 141, 143
Wicherts, J. M. 70, 74, 75, 77
Wikipedia 11, 84, 135
Wilson, L. 59
writing: advice for research reports 91–92; recommendations on 90–91

Yellowstone National Park 5, 6
Yellowstone National Park Protection Act of 1889 148

For Product Safety Concerns and Information please contact our EU representative GPSR@taylorandfrancis.com
Taylor & Francis Verlag GmbH, Kaufingerstraße 24, 80331 München, Germany

www.ingramcontent.com/pod-product-compliance
Ingram Content Group UK Ltd.
Pitfield, Milton Keynes, MK11 3LW, UK
UKHW021448080625
459435UK00012B/407

* 9 7 8 0 3 6 7 4 6 4 5 9 2 *

management-oriented and production-centered are considered to have a scientific management culture.

Fayol (1949) then created administrative science, later known as classical management, which still informs much of our organizational operations today. Fayol's approach to management, which is the creation of culture, included the following four elements: 1) structure, 2) power, 3) reward, and 4) attitude. His approach not only includes hierarchy and centralized decision making but also began to consider the employee's feelings through reward and attitude. This recognized the power that employees have within organizations and the importance of keeping employees happy in order to retain them. Many of Fayol's principles gave way to more modern management approaches such as human resources, human relations, and Theory Y (Eisenberg et al., 2017).

Today, organizational culture remains fluid and evolving, particularly with regard to the influence of technology and the changing ways we work including time and location. Similar to Fayol, we now understand organizational culture as having four elements. The four elements are: 1) strategy, 2) structure, 3) people, and 4) processes (Sanchez, 2011). Strategy is necessary to articulate how resources will be used and applied to help fulfill the mission of an organization. The structure then determines the positioning and distribution of resources that the organization needs to carry out the strategy. People, of course, are required to execute the strategy and create and maintain the structure. Finally, processes are the ways that tasks are implemented to create the functionality of the organization. When these four elements are in balance, Sanchez (2011) argues that a harmonious organizational culture, one that is positive and supportive of the organizational mission, exists.

5.3 Previous Research and Theory

Researchers have studied organizational culture so they can better understand various components of the workplace, including leadership; recruitment and retention of employees; change management; and job satisfaction. What follows is a brief overview of the literature in each of these areas. Before diving into the empirical literature, it is important to note that organizational culture lacks a central theory, and research typically is informed by multiple theoretical perspectives that can be successfully studied with either and both qualitative and quantitative methods.

Leadership

Leadership, similar to organizational culture, is a robust area of inquiry that draws upon the knowledge of other disciplines such as psychology, sociology, anthropology, business, and communication. Leadership and culture are correlated as they relate to the workplace. A poor leader can drive employees and customers away, whereas an inspiring and motivational leader can not only retain employees but also grow the organization internally and externally. As argued by Schein (2010), leadership and culture are fundamentally intertwined in three ways: "1) [that] leaders as entrepreneurs are the main architects of culture, 2) that after cultures are formed, they influence what kind of leadership is possible, and 3) that if elements of the culture become dysfunctional, leadership can and must do something to speed up cultural change" (p. xi).

Employee Recruitment and Retention

Particularly when examining organizational culture from a generational perspective, culture operates symbiotically with employee recruitment and retention. As discussed in the previous chapter, employees have expectations about what they want from organizations. This concept has been referred to as "anticipatory socialization" in organizational literature and focuses on forming expectations among potential employees (Dubinsky, Howell, Ingram, & Bellenger, 1986). The anticipatory socialization phase of job recruitment helps establish career expectations via realism and congruence. Realism is the degree to which recruits have a complete and accurate notion of what life is really like at an organization (Dubinsky et al., 1986). Congruence is the degree to which an organization's resources and demands, as well as a job candidate's needs and skills, are compatible (Dubinsky et al., 1986). Both realism and congruence have been shown to affect job satisfaction and turnover rates, because when the career related expectations of employees are not met, they find new employment (Porter, Lawler, & Hackman, 1975). The objective of anticipatory socialization is to influence the formation of job seekers' expectations and to reduce the likelihood of unmet expectations in the future (Wanous, 1977). This relates to organizational culture because it is during the anticipatory socialization process that prospective employees learn and assess the culture of an organization and create expectations about the working environment and potential fit.

Change Management

As we have all experienced in the last few years, change is constant, rapid, and can create uncertainty among people and within organizations. As the world around us changes, organizations are forced to either keep up or shut down, so not surprisingly, organizations work hard to remain operational which creates changes in communication and culture. As Pettigrew (1985) explains, changes within an organization are a response to business and economic events based on managerial perception, choice, and action. Previous research has already demonstrated that organizational culture plays a vital role in change management (Ahmed, 1998; DeLisi, 1990; Lorenzo, 1998; Pool, 2000; Schneider & Brief, 1996; Silvester & Anderson, 1999). This research helps to understand the association between organizational culture and attitudes toward organizational change—in other words, organizations whose organizational culture is evaluated positively are more likely to have employees with positive attitudes toward organizational change (Rashid, Sambasivan, & Rahman, 2004). Similarly, a qualitative study indicated that when employees have values that are congruent with those of the organization, they react more positively to change. Furthermore, because change can be an emotional experience, when organizations treat employees with respect during a period of change, those people become more engaged with the change (Smollan & Sayers, 2009).

Job Satisfaction

Another major area of inquiry related to organizational culture is job satisfaction. It is sensible to reason that when someone is satisfied with the organizational culture, they are more likely to also be satisfied with their job. However, as research indicates, things are not always this plain and simple. Much like organizational culture, job satisfaction is a multidimensional concept that is influenced by various internal and external factors. Job satisfaction depends on many organizational variables such as size, structure, salary, working conditions, and leadership, all of which constitute organizational culture. Early research about the relationship between organizational culture and job satisfaction indicated that job satisfaction increases as people progress to higher job levels and is based on a productive working environment (Corbin, 1977; Schneider & Snyder, 1975). Later, this research was extended by Sempane Rieger and Roodt (2002) using job satisfaction as a way to predict perceptions of organizational culture. Collectively, research in the areas of job satisfaction and organizational culture remains ongoing but demonstrates that there is a relationship between the two that requires attention from organizational leaders.

5.4 Generational Perspectives on Organizational Culture

While organizational culture has been studied in various contexts as discussed, organizational culture is also influenced by generational expectations about the workplace. At least in mass media, the multigenerational workplace is often discussed, sometimes only for comedic relief, showing that Millennials never want to talk on the phone, that Boomers cannot keep up with technological advances, and that Gen X is just merely there. This section will break down the different generational perspectives about organizational culture that exist before presenting our multigenerational data about organizational culture.

Millennials

Millennials are very outspoken about their cultural desires when entering the workforce. For instance, they have inspired the inclusion of many previously unconventional concepts, such as nap pods, pet-friendly workplaces, and co-working spaces. One of the most desired variables of a Millennial-friendly organizational culture is flexibility (Rawlins, Indvik, & Johnson, 2008) and the ability to work flexible hours (Brack & Kelly, 2012). This does not imply that Millennials do not want to work, as often reported in mass media outlets, but means that Millennials want to work on their own terms, when and where they want, to accommodate their lifestyles.

Speaking of lifestyle, work/life balance is another important component of organizational culture for Millennials. This could explain why Rawlins et al. (2008) found that Millennials want to be able to manage their personal lives while at work, if needed. A workplace that has a culture of "being seen" and logging long hours regardless of actual productivity is not the right fit for this generation. In fact, it is estimated that Millennials would be willing to give up $7,600 in salary every year to work in a desired environment (Chew, 2016). This is a major shift from previous generations who viewed work as a way to live.

Millennials value several elements of organizational culture, including corporate social responsibility; diversity and inclusion; work/life balance; results-oriented discussions through feedback and growth; and purposeful engagement (Alton, 2017). Millennials want to work for organizations that share and support their own beliefs, particularly with regard to environmental and social issues. Therefore, when organizations offer corporate social responsibility programs like recycling, carpools, and available time-off to volunteer, Millennials are more likely to apply for jobs there. This is related to Millennials politically independent viewpoints, which influence their need for diversity and inclusion in the workplace. Organizations that demonstrate support for diverse groups, by supporting gay pride month or International Women's Day for instance, earn stronger recruitment and retention efforts among Millennials.

Millennials are goal-oriented, and they thrive in cultures that give feedback and promote growth. On-the-job learning opportunities, employee resource groups, and both formal and informal mentoring and evaluations are attractive to Millennials. While the list of examples could go on and on, the bigger point is that when Millennials identify with an organization's culture, they are more likely to apply for positions and remain employed there.

Generation X

After Millennials, Generation X is the second largest generation present in today's workforce. This means that they not only have a lot of control over organizational culture but also a lot of experience to determine different variables present in an ideal organizational culture. Unlike Baby Boomers, Gen X has more experience with technology in the workplace, but they still have less digital wisdom than Millennials. This is helpful because this cohort is not afraid of technological changes, especially at work, since they have experienced rapid changes throughout their lives (Allen, 2017).

Gen X'ers are known for their high-quality work output and dedication to the work itself. Their values align with the core foundation of organizational culture, which is to work toward a shared goal or mission. In the workplace, Gen X'ers make great mentors to Millennials, while also being role models as leaders and helping to shape the future leaders within the organization (Allen, 2017). These factors contribute greatly to organizational culture. Much like Millennials, Gen X also desires a flexible organizational culture and likes to have strict work-life balance boundaries. Additionally, because Gen X has faced more underemployment than Boomers and was the first generation to enter the workforce with large debts from education, they are motivated by salaries in the workplace rather than, or instead of, other perks that can be appealing to Millennials (Mulvanity, 2001). Finally, because Boomers are choosing to remain in the workforce beyond the once standard retirement age of 65, many Gen X'ers view their opportunities for advancement as grim. Therefore, many Gen X'ers make several lateral moves among organizations to soak up as much knowledge and money as they can before moving on to a new job (Mulvanity, 2001). That's why considering how organizational culture influences Gen X is important for retaining this generation.

Some of the defining workplace characteristics of Gen X'ers include the need for appreciation, development, involvement, recognition, direct communication, and sincerity (Muchnick, 1996; Raines, 1997). In order to keep Gen X'ers happy at work, there cannot be any micromanagement, and the workplace needs to feel flexible and fun to them, elements that are inherent in organizational culture. This is also a generation that is strongly against "paying their dues" and wants recognition to be based solely on merit. Due to their needs for freedom, work-life balance, and flexibility, the best organizational culture fit for Gen X'ers is one where empathy is practiced. Leadership should understand that work does not occur in a vacuum and requires support for family, health, and diversity needs.

Baby Boomers

One of the biggest concerns and points of interest with Baby Boomers in the workplace is their use of technology. It is a misconception that Boomers are unable or unwilling to learn and adapt to new technologies. The reality is that during their tenure in the workforce, they have experienced a great amount of change, all of which have been implemented for regular use. For instance, this generation went from a world relying on phone calls and fax machines to the takeover of email and video conferencing. Baby Boomers can and will adapt to technology as long as the new tools are developed to make their work and lives easier (Marx, n.d.).

Baby Boomers are also in the unique position to both be a mentor and mentee within an organization. Since Boomers are less concerned right now with upward mobility, compared to Millennials, they are in an excellent position to serve as a mentor and help Gen X and Millennials within the organization. This helps make Baby Boomers feel appreciated and valued, which they like (Marx, n.d.). In some organizations, the practice of two-way mentoring is known as "transferring tribal knowledge" and has been used in organizations such as General Electric, Estee Lauder, and Saint-Gobain North America (Altany, 2019).

5.5 Organizational Culture and Generational Differences: Our Data

Perceptions about the strength of different dimensions of organizational culture were measured with the Organizational Culture Survey (Glaser, Zamanou, & Hacker, 1987). In addition to the overall scale, six sub-dimensions are included in the measure: Teamwork, Morale, Information Flow, Involvement, Supervision, and Meetings. Greater scores on this 5-point scale indicate greater endorsement of that sub-dimension of culture being present in their current organization. Overall, the average scores across all subscales were clustered above the midpoint, ranging from 3.33 to 3.66. Scores per generational group are presented in Table 5.1.

As the table demonstrates, there are differences by generation in only one of the six subscales and main scale: Morale. Gen X has a significantly lower morale score than both Baby Boomers and Millennials, who do not differ from one another. Gen X's dissatisfaction in the workplace has been demonstrated in other areas, as seen throughout other chapters in this volume. Similar outlooks on organizational culture represent an agreement in criteria for evaluating each of these facets of culture. The lack of differences in perceptions of culture is an interesting one.

The data from this study present interesting avenues for both practice within organizations and future research. What this demonstrates is that

Table 5.1 Generational Differences in the Organizational Culture Survey and Subscales

Measure	*Baby Boomers* M(SD)	*Gen X* M(SD)	*Millennials* M(SD)	F *(2, 1147)*	*eta²*
Organizational Culture Survey	3.67(.77)	3.48(.85)	3.58(.78)	2.290	.00
Teamwork	3.69(.80)	3.55(.83)	3.60(.79)	1.046	.00
Morale	3.66(.90)[a]	**3.38(.1.03)**[b]	3.61(.94)[a]	5.747*	.01
Information Flow	3.56(.87)	3.44(.93)	3.52(.87)	.964	.00
Involvement	3.51(.99)	3.29(1.11)	3.39(.98)	1.653	.00
Supervision	3.75(.95)	3.62(.96)	3.67(.89)	.751	.00
Meetings	3.41(.91)	3.31(.95)	3.33(.94)	.385	.00

$^{**}p < .001$, $^{*}p < .05$.

[a,b]Significant differences between groups as determined by Tukey HSD post hocs. Scores that significantly differ from the other two scores are bolded.

Note: Morale differences between Gen X and Baby Boomers $p = .051$.

many organizations are already successfully retaining a multigenerational workforce, which is great news. This also demonstrates that while each generation has its own "wish list" of an ideal workplace, when various components of a workplace are simultaneously working in harmony, it creates an overall culture of satisfaction. For instance, meetings can help people feel involved, supervised, and part of a team.

It is important to note that this scale does not take generational desires into account, thereby making it more objective, but also less specific to generational differences that could still be present. It would be difficult for any organization to operate without any of the measures included here. That does not mean that these are the only measures related to culture. Further, this scale does not allow the inference that each generation is satisfied with the organizational culture.

5.6 Best Practices for Creating an Age-Inclusive Organizational Culture

Clarity and Communication

Clarifying the mission, values, and operations of the organization creates an organizational culture. Therefore, organizations should work hard to create clarity around these things then implement their communication about these concepts to existing employees as well as to future employees during recruitment. Working toward a shared goal can positively influence morale, teamwork, and involvement, all of which employees use to assess culture, as our data demonstrate. Moreover, knowing that generations take the values and mission of organizations seriously throughout their job searching process (see Chapter 4 for additional information), organizations should not let the opportunity to showcase their culture during recruitment pass by. Finally, communicating and showing how the organization actively works to fulfill its mission during the recruitment process helps with anticipatory socialization.

Assess Your Culture

Both organizational leaders and employees can assess the company culture that is present within the workplace. Doing this can help determine whether or not the actual culture is in alignment with the desired culture. Many of the steps for assessing organizational culture are similar to those outlined in Chapter 3 about internal communication. Beginning with a survey of all employees, including leaders, to understand how they perceive the culture is vital. This will encourage involvement for all employees, allowing them to share their thoughts, as well as provide information that is quantifiable, which can later be used to refine and refresh policies and operating procedures (Sanchez, 2011). An organizational culture measurement study can also be done using qualitative methods such as employee interviews or focus groups and/or observation. The results from a cultural study should then be examined against the desired culture and assessed with input from leadership so that there is a collective understanding of the current culture, ways to maintain it, and how to improve it in the future (Sanchez, 2011).

5.7 Conclusion

While the data from our study are consistent with that of other empirical research which does not show a major generational difference in perceptions of organizational culture, this is still an important finding. The information presented in this chapter coupled with our data shows that culture is holistic and communal. Organizational culture, when established well, can transcend age and experience levels to create an inclusive environment where everyone is happy, motivated, and encouraged to work toward the same goals guided by one central mission.

References

Ahmed, P. K. (1998). Culture and climate for innovation. *European Journal of Innovation Management*, *1*, 30–43.

Allen, D. (2017). The merging of Gen X and Millennial cultures in the workplace. *ATD: Association for Talent Development*. Retrieved from https://www.td.org/insights/the-merging-of-gen-x-and-millennial-cultures-in-the-workplace

Altany, K. (2019). Baby Boomers vs. Millennials: Merging culture. Retrieved from https://www.industryweek.com/talent/article/22028374/baby-boomers-vs-millennials-merging-cultures

Alton, L. (2017). How millennials are reshaping what's important in corporate culture. *Forbes*. Retrieved from https://www.forbes.com/sites/larryalton/2017/06/20/how-millennials-are-reshaping-whats-important-in-corporate-culture/#f1da7222dfb8

Barney, J. (1986). Organizational culture: Can it be a source of sustained competitive advantage? *Academy of Management*, *11*, 656–665.

Brack, J., & Kelly, K. (2012). Maximizing Millennials in the workplace. *UNC Kenan-Flagler Business School*. Retrieved from https://www.kenan-flagler.unc.edu/executive-development/custom-programs/~/media/files/documents/executive-development/maximizing-millennials-in-the-workplace.pdf

Braddy, P., Meade, A., & Kroustalis, C. (2006). Organizational recruitment website effects on viewers' perceptions of organizational culture. *Journal of Business and Psychology*, *20*, 525–543.

Chew, J. (2016). Why Millennials would take a $7,600 pay cut for a new job. *Fortune.* Retrieved from http://fortune.com/2016/04/08/fidelity-millennial-study-career/

Corbin, L. (1977). Productivity and job satisfaction in research and development: Associated individual and supervisory variables. *Airforce Institute of Technology*, *3*.

Deal, T., & Kennedy, A. (1982). *Corporate cultures.* Addison-Wesley.

DeLisi, P. S. (1990). Lessons from the steel axe: Culture, technology, and organizational change. *Sloan Management Review*, *32*, 83–93.

Dubinsky, A., Howell, R., Ingram, T., & Bellenger, D. (1986). Salesforce socialization. *Journal of Marketing*, *50*, 192–207.

Eisenberg, E., Trethewey, A., LeGreco, M., & Goodall, H. (2017). *Organizational communication: Balancing creativity and constraint* (8th ed.). Bedford St. Martin's.

Fayol, H. (1949). *General and industrial management.* Pitman.

Glaser, S. R., Zamanou, S., & Hacker, K. (1987). Measuring and interpreting organizational culture. *Management Communication Quarterly*, *1*, 173–198.

Hofstede, G. (1994). *Uncommon sense about organizations: Case studies and field observations.* Sage.

Lorenzo, A. (1998). A framework for fundamental change: Context, criteria, and culture. *Community College, Journal of Research & Practice*, *22*, 335–348.

Marx, L. (n.d.). How to create a company culture that Baby-Boomers, Millennials, and Gen Z employees can thrive in. Retrieved from: https://www.urbanbound.com/blog/how-to-create-a-company-culture-that-baby-boomers-millennials-and-gen-z-employees-can-thrive-in

Muchnick, M. (1996). *Naked management: Bare essentials for motivating the X-Generation at work.* St. Lucie Press.

Mulvanity, E. (2001). Generation X in the workplace: Age diversity issues in project teams. *Project Management Institute.* Retrieved from https://www.pmi.org/learning/library/generation-x-workplace-age-diversity-style-7904

Pettigrew, A. (1985). *The awakening giant: Continuity and change in imperial chemical industries.* Blackwell.

Pool, S. (2000). Organizational culture and its relationship between job tension in measuring outcomes among business executives. *Journal of Management Development*, *19*, 32–49.

Porter, L., Lawler, E., & Hackman, J. (1975). *Behavior in organizations.* McGraw Hill.

Raines, C. (1997). *Beyond Generation X.* Crisp Publications.

Rashid, M., Sambasivan, M., & Rahman, A. (2004). The influence of organizational culture on attitudes toward organizational change. *Leadership & Organizational Development Journal*, *25*, 161–179.

Rawlins, C., Indvik, J., & Johnson, P. (2008). Understanding the new generation: What the millennial cohort absolutely positively must have at work. *Journal of Organizational Culture, Communications and Conflict*, *12*, 1–8.

Sanchez, P. (2011). Organizational culture. In T. Gillis (Ed.), *The IABC handbook of organizational communication* (pp. 28–40). Jossey-Bass.

Schein, E. (2010). *Organizational Culture and Leadership* (4th ed.). Josey-Bass.

Schneider, B., & Brief, A. (1996). Creating a climate and culture for sustainable organizational change. *Organizational Dynamics*, *24*, 7–19.

Schneider, B., & Snyder, R. (1975). Some relationship between job satisfaction and organizational climate. *Journal of Applied Psychology*, *60*, 318–328.

Sempane, M., Rieger, H., & Roodt, G. (2002). Job satisfaction in relation to organizational culture. *Journal of Industrial Psychology*, *28*, 23–30.

Sheridan, J. (1992). Organizational culture and employee retention. *Academy of Management Journal*, *35*, 1036–1056.

Silvester, J., & Anderson, N. (1999). Organizational culture change. *Journal of Occupational & Organizational Psychology*, *72*, 1–24.

Smollan, R., & Sayers, J. (2009). Organizational culture, change and emotions: A qualitative study. *Journal of Change Management*, *4*, 435–457.

Trompenaars, F., & Hampden-Turner, C. (1997). *Riding the waves of culture: Understanding diversity in global business*. McGraw Hill.

Umphress, E., Bingham, J., & Mitchell, M. (2010). Unethical behavior in the name of the company: The moderating effect of organizational identification and positive reciprocity beliefs on unethical pro-organizational behavior. *Journal of Applied Psychology*, *95*, 769–780.

Wanous, J. (1977). Organization entry: Newcomers moving from outside to inside. *Psychological Bulletin*, *84*, 601–618.

6 Organizational Identification

As part of the larger corporate structure, employees must still feel connected to the organization. This chapter will discuss how employees identify with and assimilate to the organization. We reflect on our own findings and connect these findings to previous literature surrounding organizational identification. Best practices will also be provided to help employees identify and connect to the organization as a whole.

6.1 Connecting to the Workplace

The traditional workplace involves people, processes, and plans. Because of this complexity, it is not enough to assume that people will feel connected to their organization simply by having a sense of belonging or accomplishment. Instead, several factors influence how and why employees feel like they are connected to their places of employment.

Employees tend to feel connected to their organizations when the socialization and assimilation processes have been effectively utilized. Employees, especially newly onboarded employees, must feel connected to their workgroups, other employees, and the larger organization as a whole (Morrison, 2002). Clear communication and relationship building can help build and sustain an employee's sense of belonging (Kammeyer-Mueller, Wanberg, Rubenstein, & Song, 2013).

Ultimately, the identity of the employee must be connected to the organization, and employees must be assimilated into the organization. As a result, retention and overall satisfaction may increase.

6.2 Identification

6.2.1 Previous Research and Theory

Identity manifests itself in many ways, especially in the organization. Identity, or "that which is central, distinctive, and more less enduring"

(Ashforth, 2016, p. 262), positions the identification concept as a personal distinctive that can be applied in organizations. Organizational identification includes one's linkage, or perceived link, to the organization. Cheney (1983), the original architect of the organizational identification questionnaire, believes identification is a process, specifically an active process whereby individuals link themselves to elements in that social scene. This link can lead to organizational commitment, i.e. retention (Cook and Wall, 1980). Thus, in the case of our study, an employee's central and distinct identity can be applied and developed within organizations, and, theoretically, the more an organization connects to an employee's central, distinct, and enduring identity, the more loyal an employee will remain.

Identification, while important especially as one enters an organization, is also an evolving concept. For one, scholars recognize that we continually re-negotiate our identity as it relates to an organization (Brown, 2017; Haslam, 2004). Further, as Kanungo (1982a) illustrates, job identification, rather than organizational commitment, can also include how committed one is to his or her professional position. In addition, identity to a position or an organization can manifest itself through concern with one's present position (Paullay, Alliger, & Stone-Romero, 1994), self-esteem related to job performance (Lodahl & Kejner, 1965), and genuine care and concern for one's work (Kanungo, 1982b). These elements, as one can imagine, tend to result in greater organizational effectiveness (Uygur & Kilic, 2009).

From a generational perspective, identification is felt differently depending on the general age group. And, ironically, for organizations to survive and thrive with a multigenerational workforce, organizations must deal not only with the social identity of their workers and the tendency to categorize ourselves and view other generations more positively or negatively depending on our perceptions (Ho & Yeung, 2020) but also with how this social identity connects to and infiltrates the workplace.

Generally, Baby Boomers, those affectionately referred to as career loyalists (Singh & Gupta, 2015), tend to find an identity in their jobs and are more likely to show a favorable attitude toward their job (Ng & Feldman, 2010). Historically, older workers view their jobs in a more positive light (Carstensen, 1991) and have received more gratification from the identification they find in their jobs (Wright & Hamilton, 1978). Generation X, those who tend to be more independent and self-reliant, are viewed as less loyal than their Boomer coworkers and Boomer bosses (Rottier, 2001) although a study by Davis, Pawlowski, & Houston (2016) reveals that Boomers and Generation X tend to be more alike in their work and job involvement as well as organizational and professional commitment. Compared to Baby Boomers and Gen-Xers, Millennials tend to be more concerned, overall, with their identity and organizational identifications. Millennials are usually more attracted to and identify more uniquely with organizations and institutions where there the system is equitable. To put it another way, despite public opinion that would decry millennials as disloyal, millennials tend to be loyal to those organizations which are loyal to them (Hershatter & Epstein, 2010). All of this impacts how one's identity is connected to and developed within the organization, and because of generational differences related to organizational identification, job identification, workplaces—regardless of the industry—would do well to establish clearer opportunities for their workers to connect to the organization at large. As workers strive to identify with their organizations, the topic needs continual study, and it is important to consider the extent to which individuals identify with the organization. And, while worker identification is an important generational consideration, how workers assimilate in their organizations and how organizations retain their employees are also important to note for those interested in organizational dynamics.

6.2.2 Organizational Identification and Generational Difference: Our Data

Cheney's (1983) Organizational Identification Scale offers a unidimensional approach to assessing the degree to which individuals identify with their organization. Across all three generations, this scale reported excellent reliability (Cronbach's alpha = .943) and mean just above its midpoint ($M = 3.76$ ($SD = 1.24$) on a 7-point scale.

A look at differences in means across the generational groups show small, yet marginally significant differences between groups. An ANOVA showed overall small, significant differences, $F(2, 1148) = 3.356$, partial eta$^2 = .006$, $p < .05$ across the three groups. Millennials reported the greatest identification with their organization ($M = 4.31$, $SD = 1.20$), followed by Baby Boomers ($M = 4.11$, $SD = 1.31$) and Gen X ($M = 4.10$, $SD = 1.31$). A Tukey post hoc test revealed the difference between Gen X and Millennials neared significance ($p = .053$).

The data from our study show that individuals' identification with their organization does not differ to a large degree by their generation. However, we see a small, and nearly significant difference: Millennials have the greatest identification with their organization as measured here. This is in line with previous research.

6.3 Assimilation

6.3.1 Previous Research and Theory

Through organizational assimilation, employees learn about the organization as a whole and learn about the organization's members, policies, procedures, culture, and other crucial aspects of the workplace that employees need to grasp (Croucher, Zeng, & Kassing, 2016). In 2003, Myers and Oetzel developed six dimensions of the organizational assimilation process. Their dimensions include *familiarity* with others, specifically the ability to develop and build relationships; *acculturation*, or the process of learning about organizational norms; *recognition*, being recognized as a valuable member of the organization by the organization; *involvement*, ways to contribute to the organization; *job competency*, or one's actual job performance; and *role negotiation*, or the process an employee goes through to negotiate their actual place or expectations within an organization. The sections that follow explore these dimensions from a generational perspective.

Familiarity

As employees assimilate into the organization, they can develop a familiarity with those other coworkers, especially supervisors. Ironically, Millennials and Gen Xers share similar perspectives on work, and their understanding tends to run counter to a Boomer perspective. Generally, for Millennials and X-ers alike, their identity is not as focused on their job (Marston, 2007). This can become a point of contention between generations in organizations and can, therefore, negatively impact how relationships are developed (Raines, 2003). As these relationships form—especially now as Boomers are still primarily supervisors—generally, in organizational contexts, these relationships become even more delicate, and familiarity becomes even more important. As Jokisaari and Nurmi (2009) found, strong relationships with supervisors help develop long-term job satisfaction for millennials. This means organizations should enhance opportunities for cross-generational relationship building. However, this may be easier than it sounds because several factors influence how we come to build and sustain relationships at work.

Obviously, trust can influence how individuals develop and build relationships, especially with supervisors. As members of different organizations strive to build effective relationships, age diversity can continually get in the way as intergenerational differences can influence the trust one has for members of their team or their supervisors/subordinates (Williams, 2016). It is important, then, to step back and recognize that in order for generational assimilation to occur successfully, there must be a foundation of trust. In organizations where communication is a prominent mechanism for culture, understanding, and general functionality, clear expectations for workplace relationships are necessary. For one, Millennials tend to expect closer relationships and greater transparency from their supervisors (Society for Human Resource Management, 2009). And it is likely that Generation Z supervisor expectations will follow a similar, or even more intense, trajectory (Goh & Lee, 2018). Familiarity and relationship building, then, become a crucial factor for assimilating workers, especially younger employees, into the organization.

Acculturation

Familiarity with members of the organization should be preceded by an acculturation process whereby the employee is immersed in an understanding of organizational norms and culture. Myers and Oetzel (2003) also focus not just on learning behavioral norms but on avoiding those actions that may go against or break organizational norms. While organizational norms look differently depending on the organization, one generational impact on organizational acculturation was the relationship between supervisor and employee. In most cases, for instance, Baby Boomers were probably taught to accept direction, criticism, feedback, etc. from a supervisor and to subsequently not question the comments (Stewart, Oliver, Cravens, & Oishi, 2017). For Millennials, generally, this organizational norm is archaic. Interestingly, Stewart et al. (2017) also argue that millennials, compared to previous generations, carry an expectation that the workplace will cater to or accommodate their needs. This understanding, coupled with the notion that organizational norms may be of less importance to millennials, means organizations should consider how to approach acculturation as a valuable standard rather than a burdensome reality.

Recognition

Generally, all employees want and need recognition in the workplace. While the type and amount of recognition may be preferential, generations across the spectrum want to know that they have performed at or above expectations (Van Dyke & Ryan, 2013). Recognition, the understanding that an employee is of value to their organization, can manifest itself in increased wages or even through additional perks like flexible hours or remote work (Meister & Willyerd, 2010). Ironically, rewards like remuneration and benefits and a positive working environment are more important to the younger employees and tend to reduce as employees get older (Close & Martins, 2015). Twenge and Campbell (2010) also argue that Millennials, compared to Baby Boomers, are more interested in extrinsic rewards but less interested compared to Generation X employees. Millennials also prefer non-traditional or non-material rewards more than their generational counterparts do and greatly appreciate learning and development and immediate feedback from supervisors (Hewlett, Sherbin, & Sumberg, 2009). These generational differences force managers to recognize the employees as individuals, understanding that a one-size-fits all reward system is outdated.

Involvement

Employee involvement in the organization, or the actual ways one contributes to the workplace, can manifest itself in a variety of ways. Boomers, for instance, can be seen by some as problematic not because of their contributions but rather because of their inability or unwillingness to adapt to a changing workplace (Bosco & Harvey, 2013). In terms of asset accumulation, Boomers are retiring with more wealth than any previous generation; while this is not a direct indicator of workplace productivity, it can shed some light on Boomer organizational and economic impact (Roberts, 2011). In a 2005 study, specifically looking at the differences in productivity between Boomers and Gen Xers, Appelbaum, Serena, & Shapiro found that older workers are more productive than younger workers due to their experience and knowledge. Long held notions that younger employees who are first starting out tend to produce more immediate results may not be entirely true. Instead, it is important to recognize that employee involvement and contributions may be linked as much to experience as anything else. However, Martin (2005) indicates that Generation Y, otherwise known as Millennials, if led by the right type of manager, have the potential to be the highest performing generation in history. This holistic understanding, that Generation Y has immense productivity potential, serves researchers well as they consider the multigenerational workplace. There are no substitutes for experience, but because of their technology acumen, entrepreneurial spirit, and achievement personality, Millennials may become dominant producers in organizational contexts.

Job Competency

In a similar vein to involvement, job competency involves one's actual job performance. While not a generational difference per se, it is important to note that historically researchers would agree that there is little to no relationship between age and job performance (Salthouse & Maurer, 1996; Warr, 1994). Job performance or competency, like involvement, can be influenced by experience and accumulated job knowledge, thus making someone who has been in a position longer potentially more efficient. Further, job performance may also filter down to different generations in the workplace as younger generations may see increased productivity because of the transferable experience of older generations through mentorship or collaborative projects (Waljee, Vineet, & Saint, 2020). As workers navigate their actual performance on the job, generational differences can be useful for considering improved productivity related to collaboration and individual duties.

Role Negotiation

The final dimension for Myers and Oetzel (2003), role negotiation, explores the process an employee goes through to negotiate their actual place or expectations within an organization. Practically speaking, role negotiation is how one perceives he or she fits within an organization. While other generations are not immune to the challenges associated with role negotiation in organizations, Millennials, because of stereotypes about them as well as their own expectations, may struggle with organizational socialization (Marston, 2007). Internal role negotiation is also determined by membership negotiation, where current organizational members decide who may, or may not, suffice as an appropriate organizational fit (Slaughter & Zickar, 2006). Those just entering positions also participate in this negotiation process (Scott & Myers, 2010). Ironically, Millennial roles will be influenced by their own perceptions, and these expectations may also impact the role(s) of others within the organization (Myers & Sadaghiani, 2010). How these negotiated roles within an organization, especially with the incoming Generation Z cohort filtering in slowly, affect long-term organizational culture is yet to be determined.

6.3.1 Organizational Assimilation and Generational Difference: Our Data

Meyers and Oetzel's (2003) Organizational Assimilation Index demonstrated excellent reliability (Cronbach's alpha = .938) and a mean score

Table 6.1 Generational Differences in the Organizational Assimilation Index and Subscales

Measure	*Baby Boomers* M(SD)	*Gen X* M(SD)	*Millennials* M(SD)	F (2, 1147)	*eta*^{2}
Organizational Assimilation Index	4.03(.68)[a]	3.82(.80)[a]	**3.76(.73)[b]**	5.77*	.01
Supervisor Familiarity	3.83(.90)	3.72(1.03)	3.75(.97)	.425	.00
Acculturation	4.47(.72)[a]	4.24(.89)[a]	**4.13(.83)[b]**	7.80**	.01
Recognition	**4.17(.92)[b]**	3.83(1.14)[a]	3.83(.99)[a]	4.860*	.01
Involvement	3.57(1.04)	3.37(1.12)	3.41(1.05)	1.20	.00
Job Competency	4.18(.72)[a]	4.03(.80)[a]	**3.82(.80)[b]**	13.574**	.02
Role Negotiation	**3.97(.92)[a]**	3.84(1.00)[b]	3.64((1.00)[b]	7.325**	.01

*$p < .001$, **$p < .05$.

[a,b]Significant differences between groups as determined by Tukey HSD post hocs. Scores which significantly differ from the other two scores are bolded.

well above the midpoint across all individuals (M = 380, SD = .74). Greater scores are indicative of greater assimilation and its subdimensions. Results are presented below in Table 6.1. Millennials scored significantly lower (M = 3.76, SD = .73) than both Gen X (M = 3.82, SD = .80) and Baby Boomers (M = 4.03, SD = .68), who did not significantly differ, despite the differences in mean scores.

The pattern of Millennials scoring significantly lower is a pattern found in the subdimensions of the scale as well. In two of the subscales, Millennials scored significantly lower than both Gen X and Baby Boomers (acculturation, job competency); in another two subscales, both Millennials and Gen X scored significantly lower than Baby Boomers (recognition, role negotiation), and there were no differences in the remaining two subscales (supervisor familiarity, involvement). While these differences are small in effect size, the pattern observed is relatively consistent: Millennials demonstrate lower scores across the subdimensions of assimilation, Baby Boomers demonstrate greater scores, and Gen X falls in between. It should be noted that across all participants, scores on organizational identification and organizational assimilation share a r = .306, $p < .01$. This moderate-size correlation speaks to the relatedness of the two concepts.

6.4 Retention

Optimizing talent, specifically hiring and retaining, is a significant challenge facing modern organizations (Clare, 2009). The elements listed above related to assimilation must be negotiated in an effective manner for employees to stay at organizations. At times, leaving a company cannot be helped. However, there are certainly instances when workers will leave an organization because they found a better opportunity, have not assimilated themselves into the broader culture, or believe they are not being used effectively in their current role. Baby Boomers still have a desire to work and participate actively in organizations (Salb, 2015). Because many Boomers are career loyalists, retention was not a necessary focus. Retention, partly because Generation X-ers were more inclined to leave their positions for something else, was more important with those who came after Boomers. Those employing Generation X need to offer variety, simulation, and constant change to keep workers engaged (Jurkiewicz, 2000). Jurkiewicz (2000) also notes that culture is of primary importance for X-ers. This trend has continued, and today employee retention is a renewed area of study.

Generational workplace preferences do vary, although some of the foundational desires of employees transcend generational demographics. For one, Eversole, Venneberg, and Crowder (2012) emphasize the importance of organizational flexibility to retain workers across the generational spectrum. Pregnolato, Bussin, and Schlecter (2017), when evaluating reward preferences of different generations, reveal that financial rewards, including benefits; performance and recognition; remuneration; and career; as well as career advancement; learning; and work-life balance are all elements that can help organizations retain talented employees. Generation Z, those post-millennial workers, likewise appreciate a career path, flexible work conditions, and transparency (Goh & Okumus, 2020). Despite their simplicity, these components are difficult to achieve in organizations.

6.5 Best Practices for Helping Employees Identify with the Organization

The integration of multiple generations into the workplace can create organizational uncertainty. Organizations should strive to create environments where employees feel as though they belong and know they are valuable members. The following best practices can help employees identify with their organization, thus contributing to assimilation and retention mechanisms that transcend generational boundaries.

Best Practice 1: Build a Mission-Centric Culture

Employees, generally, respond more positively to organizations where they feel connected. This is especially true for members of Generation Z as well as Millennials. Yet, Gen X-ers and Baby Boomers also want to know that their work has value. Organizations would do well to create a mission-centric culture that is inclusive of all backgrounds and perspectives. But, ultimately, reminding employees of the mission and articulating how each individual employee "fits" into the broader work can be helpful.

Best Practice 2: Communicate Clearly

A sense of belonging does not occur naturally and, instead, must be communicated from the supervisory level and through peer relationships. Communicate expectations and explore ways to deliver messages in ways that resonate with individual employees, not just the group as a whole. As you onboard employees, you must also communicate effectively as the assimilation process starts.

Best Practice 3: Start Building Relationships Early and Often

Employees must be familiar with the organization and their role within the broader structure. This familiarity occurs over time, but you can develop mentorship relationships, especially across the generational spectrum, that help younger workers integrate, encourage older workers to leave a legacy of impact, and establish a culture of relationship development throughout the organization.

6.6 Conclusion

Our data show an interesting connection to previous research. Millennials want and need to feel connected to the institution. Assimilation, and the subsequent connection that follows, is key to establishing rapport with younger employees and can help retain and even recruit Millennial talent. It is important to continue to stress to all employees, especially younger generations, that training for the job will be provided and that the culture, as a whole, is employee friendly and generally positive.

References

Appelbaum, S. H., Serena, M., & Shapiro, B. T. (2005). Generation "X" and the Boomers: An analysis of realities and myths. *Management Research News*, *28*, 1–33. https://doi.org/10.1108/01409170510784751

Ashforth, B. E. (2016). Exploring identity and identification in organizations: Time for some course correction. *Journal of Leadership and Organizational Studies*, *23*, 361–373. doi: 10.1177/1548051816667897

Bosco, S. M., & Harvey, D. M. (2013). Generational effects on recruitment and workplace productivity. *Northeast Business and Economics Association Proceedings*, 17–20.

Brown, A. D. (2017). Identity work and organizational identification. *International Journal of Management Reviews*, *19*, 296–317. doi: 10.1111/ijmr.12152

Carstensen, L. L. (1991). Selectivity theory: Social activity in life-span context. In K. W. Sehaie (Ed.), *Annual review of gerontology and geriatrics* (pp. 195–217). Springer.

Cheney, G. (1983). On the various and changing meanings of organizational membership: Field study of organizational identification. *Communication Monographs*, *50*, 342–362.

Clare, C. (2009). Generational differences: Turning challenges into opportunities. *Journal of Property Management*, *74*, 40–41.

Close, D., & Martins, N. (2015). Generational motivation and preference for reward and recognition. *Journal of Governance and Regulation*, *4*, 259–270.

Cook, J., & Wall, T. (1980). New work attitude measures of trust, organizational commitment and personal need non-fulfillment. *Journal of Occupational Psychology*, *53*, 39–52.

Croucher, S. M., Zeng, C., & Kassing, J. (2016). Learning to contradict and standing up for the company: An exploration of the relationship between organizational dissent, organizational assimilation, and organizational reputation. *International Journal of Business Communication*, *56*, 349–367. https://doi.org/10.1177%2F2329488416633852

Davis, J. B., Pawlowski, S. D., & Houston, A. (2016). Work commitments of Baby Boomers and Gen-Xers in the IT profession: Generational differences or myth? *Journal of Computer Information Systems*, *46*, 43–49. https://doi.org/10.1080/08874417.2006.11645897

Eversole, B. A. W., Venneberg, D. L., & Crowder, C. L. (2012). Creating a flexible organizational culture to attract and retain talented workers across generations. *Advances in Developing Human Resources*, *14*, 607–625. https://doi.org/10.1177%2F1523422312455612

Goh, E., & Lee, C. (2018). A workforce to be reckoned with: The emerging pivotal Generation Z hospitality workforce. *International Journal of Hospitality Management*, *73*, 20–28. https://doi.org/10.1016/j.ijhm.2018.01.016

Goh, E., & Okumus, F. (2020). Avoiding the hospitality workforce bubble: Strategies to attract and retain Generation Z talent in the hospitality workforce. *Tourism Management Perspectives*, *33*, 1–7. https://doi.org/10.1016/j.tmp.2019.100603

Haslam, S. A. (2004). *Psychology in organizations: The social identity approach* (2nd ed.). Sage.

Hershatter, A., & Epstein, M. (2010). Millennials and the world of work: An organization and management perspective. *Journal of Business Psychology*, *25*, 211–223. doi:10.1007/s10869-010-9160-y

Hewlett, S. A., Sherbin, L., & Sumberg, K. (2009). How Gen Y & boomers will reshape your agenda. *Harvard Business Review*, *87*, 71–76.

Ho, H. C. Y., & Yeung, D. Y. (2020). Conflict between younger and older workers: An identity-based approach. *International Journal of Conflict Management*, 1–24. doi:http://dx.doi.org/10.1108/IJCMA-08-2019-0124

Jokisaari, M., & Nurmi, J. E. (2009). Change in newcomers' supervisor support and socialization outcomes after organizational entry. *Academy of Management Journal*, *52*, 527–544.

Jurkiewicz, C. L. (2000). Generation X and the public employee. *Public Personnel Management*, *29*, 55–74. https://doi.org/10.1177%2F009102600002900105

Kammeyer-Mueller, J. D., Wanberg, C. R., Rubenstein, A., & Song, Z. (2013). Support, undermining, and newcomer socialization: Fitting in during the first 90 days. *Academy of Management Journal*, *56*, 1104–1124. https://doi.org/10.5465/amj.2010.0791

Kanungo, R. N. (1982a). Measurement of job and work involvement. *Journal of Applied Psychology*, *67*, 341–349.

Kanungo, R. (1982b). *Work alienation: An integrative approach*. Wiley.

Lodahl, T. M., & Kejner, M. (1965). The definition and measurement of job involvement. *Journal of Applied Psychology*, *49*, 24–33.

Marston, C. (2007). *Motivating the what's in it for me workforce: Managing across the generational divide and increase profits*. John Wiley & Sons, Inc.

Martin, C. A. (2005). From high maintenance to high productivity: What managers need to know about Generation Y. *Industrial and Commercial Training*, *1*, 39–44. https://doi.org/10.1108/00197850510699965

Meister, J. and Willyerd, K. (2010). *The 2020 workplace*. HarperCollins.

Morrison, E. W. (2002). Newcomers' relationships: The role of social network ties during socialization. *Academy of Management Journal*, *45*, 1149–1160. https://doi.org/10.5465/3069430

Myers, K. K., & Oetzel, J. G. (2003). Exploring the dimensions of organizational assimilation: Creating and validating a measure. *Communication Quarterly*, *51*, 438–457. https://doi.org/10.1080/01463370309370166

Myers, K. K., & Sadaghiani, K. (2010). Millennials in the workplace: A communication perspective on Millennials' organizational relationships and performance. *Journal of Business and Psychology*, *25*, 225–238.

Ng, T. W. H., & Feldman, D. C. (2010), The relationships of age with job attitudes: A meta-analysis. *Personnel Psychology*, *63*(3), 677–718.

Paullay, I., Alliger, G., & Stone-Romero, E. (1994). Construct validation of two instruments designed to measure job involvement and work centrality. *Journal of Applied Psychology*, *79*, 224–228.

Pregnolato, M., Bussin, M. H. R., & Schlecter, A. F. (2017). Total rewards that retain: A study of demographic preferences. *SA Journal of Human Resource Management*, *15*, 1–10. doi: 10.4102/sajhrm.v15.804

Raines, C. (2003). *Connecting generations: The sourcebook for a new workplace*. Thompson Crisp Learning.

Roberts, K. (2011). The end of the long baby-boomer generation. *Journal of Youth Studies*, *4*, 479–497. https://doi.org/10.1080/13676261.2012.663900

Rottier, A. (2001). Gen 2001: loyalty and values, *Workforce*, *80*, 23.

Salb, D. (2015). Using technology to retain Baby Boomers in the workplace. *Computer and Information Science*, *8*, 180–185. doi:10.5539/cis.v8n3p180

Salthouse, T. A., & Maurer, T. J. (1996). Aging, job performance, and career development. In J. E. Birren, K. W. Schaie, R. P. Abeles, M. Gatz, & T. A. Salthouse (Eds.), *Handbook of the psychology of aging* (4th ed., pp. 353–365). Academic Press.

Scott, C. W., & Myers, K. K. (2010). Toward an integrative theoretical perspective of membership negotiations: Socialization, assimilation, and the duality of structure. *Communication Theory*, *30*, 79–105.

Singh, A., & Gupta, B. (2015). Job involvement, organizational commitment, professional commitment, and team commitment: A study of generational diversity. *Benchmarking: An International Journal*, *22*, 1192–1211.

Slaughter, J. E., & Zickar, M. J. (2006). A new look at the role of insiders in the newcomer socialization process. *Group & Organization Management*, *31*, 264–290.

Society for Human Resource Management. (2009). *The multigenerational workforce: Opportunity for competitive success*. http://www.shrm.org/Research/Articles/Articles/Documents/09-0027_RQ_March_2009_FINAL_noad.pdf

Stewart, J. S., Oliver, E. G., Cravens, K. S., & Oishi, S. (2017). Managing millennials: Embracing generational differences. *Business Horizons*, *60*, 45–54. https://doi.org/10.1016/j.bushor.2016.08.011

Twenge, J., & Campbell, S. (2010). Who are the Millennials? Empirical evidence for generational differences in work values, attitudes and personality. In E. Ng, S. Lyons, & L. Schweitzer (Eds.), *Managing the new workforce* (pp. 1–19). Edward Elgar Publishing.

Uygur, A., & Kilic, G. (2009). A study into organizational commitment and job involvement: An application towards the personnel in the central organization for ministry of health in Turkey. *Ozean Journal of Applied Sciences*, *2*, 113–125.

Van Dyke, M., & Ryan, M. (2013). Changing the compensation conversation and the growing utility of non-cash rewards and recognition. *Compensation and Benefits Review*,*44*, 276–279.

Williams, M. (2016). Being trusted: How team generational age diversity promotes and undermines trust in cross-boundary relationships. *Journal of Organizational Behavior*, *37*, 346–373.
Warr, P. (1994). Age and employment. In M. Dunnette, L. Hough, & H. Triandis (Eds.), *Handbook of industrial and organizational psychology* (pp. 487–550). Consulting Psychologists Press.
Waljee, J. F., Vineet, C., & Saint, S. (2020). Mentoring millennials. *JAMA*, *17*, 1716–1717. doi:10.1001/jama.2020.3085
Wright, J. D., & Hamilton, R. F. (1978). Work satisfaction and age: Some evidence for the 'job change' hypothesis. *Social Forces*, *56*, 1140–1158.

7 Communication in the Organization: Positive Communication

This chapter will explore, specifically, supportive workplace communication by focusing on mentorship and communication support. The authors will reflect on their findings and connect these findings to previous literature surrounding mentorship and supportive communication.

7.1 Supportive Communication

The workplace is unique because it is a place where different contexts of communication converge. We have to manage interpersonal relationships and stakeholder communication as well as maintain our professional self-image simultaneously. While people are familiar with the various conflicts and dilemmas that can occur due to communication – or rather miscommunications – in the workplace, there is also a vast body of research that explores the ways in which communication helps engage employees and keep them happy, supported, and satisfied in the workplace.

Public relations helps to understand communication within the organization and between organizations and an external audience (customers, community leaders, etc.). Excellence theory is one of the most foundational theories to inform public relations and gives way to understanding supportive communication. The essence of excellence theory is that the most effective communication is two-way and symmetrical (Grunig & Grunig, 2008). This means that both parties engaged in communication are speaking and listening equally. Operating with this theory in mind helps inform supportive communication in the workplace while also helping with employee trust and engagement efforts.

Supportive communication is defined as "verbal and nonverbal behavior produced with the intention of providing assistance to others perceived as needing that aid" (Burleson & MacGeorge, 2002, p. 374). Previous research helps to understand the various benefits of supportive communication, an area that has been robustly studied in health and interpersonal contexts, especially. For instance, supportive communication has been demonstrated to create and strengthen social networks, improve social experiences, and provide greater perceptions of available support (Burleson, 2009; Cohen, Mermelstein, Karmarck, & Hoberman, 1985; Shaver, 2008). Furthermore, supportive communication can decrease emotional distress and help people cope more easily (Cunningham & Barbee, 2000; Gottlieb, 1994). Research about supportive communication has also demonstrated that supportive messages are complex and that the effectiveness of messages is contingent upon various factors, including the message source, the receiver of the message, and the interactional context (Bodie & Burleson, 2008; Cutrona, Cohen, & Igram, 1990).

Burleson (2009) outlines features of supportive messages, which apply to both professional and interpersonal interactions. The first feature is the message content, which includes verbal and nonverbal communication. The verbal content is what the person seeking the support actually says to solicit the need for support from someone else and the type of support offered through the verbal message. Different types of support include emotional support, informational support or advice, esteem support, and/or social network support. The nonverbal elements of a message can include the length of the message, the timing or placement of the message within an interaction, how the message is shared, and the number of appeals or statements in the message (Feng, 2009; Jacobson, 1986; Neff & Karney, 2005).

The second feature of supportive communication is the relationship that exists between the person needing support and the person providing the support. Women tend to be viewed as more supportive than men, and messages provided by women are evaluated as more supportive than those typically provided by men, even when the content of the message is the same (Glynn, Christenfeld, & Gerin, 1999; Uno, Uchino, & Smith, 2002). The credibility of the helper is also known to influence supportive communication messages, as well as the quality of the relationship between the two parties. When the person receiving support feels close to the one providing support, messages are evaluated more favorably (Clark et al., 1998).

The final two features include the context and the recipient. Features of the context refer to the physical setting, the medium of communication, and the problem situation that makes supportive communication a relevant activity (Burleson, 2009). For instance, unsolicited messages are viewed unfavorably, and the quality of support is more impactful when it is related to a serious issue (Burleson, 2009; MacGeorge, Feng, & Thompson, 2008). Finally, features of the recipient matter, making this timely for consideration within a multigenerational workforce. Demographic characteristics, personality composition, and cognitive attributes influence responses to supportive communication (Burleson, 2009).

7.2 Supportive Communication in the Workplace

The workplace creates endless opportunities for uncertainty, which creates the need for support. The workplace can also be a stressful environment, which furthers our need and desire for support. This is why people so often turn to their coworkers for post-work happy hours, sessions where everyone can vent their frustrations, and interoffice relationships, all of which can be considered support, albeit not always healthy support. Supportive communication in the workplace seeks to help manage and reduce uncertainty (Mikkola, 2019). The benefits of supportive communication at work are plentiful, including increases in productivity, motivation, job satisfaction, and commitment. Supportive workplace relationships can also help with problem-solving, decision making, and learning.

Mikkola (2019) outlines two types of support present in the workplace: emotional and informational support. Emotional support is what enables employees to become friends, share frustrations, and celebrate each other's accomplishments. Emotional support is used to help gain psychological distance from the emotions that can erupt during stressful and uncertain workplace situations. Through emotional support, employees help each other lessen the amount of emotional distress one or both of them may be experiencing (Burleson, 2003). A central feature of emotional support is legitimizing the feelings of the person seeking support through verbal and nonverbal communication. Informational support, on the other hand, provides relevant information to reduce uncertainty as a form of support (Brashers, 2001). Informational support is best suited to situations that are problem and solution oriented, whereas many emotional support situations cannot be solved through information.

Supportive communication and social support provide various positive outcomes in the workplace, as evidenced by previous research. For example, social support promotes high-quality performance through increasing emotional affirmation and strengthening the capacity for collective problem-solving (Park, Wilson, & Lee, 2004). Social support can also boost engagement among both employees and leaders alike, which is an antecedent to organizational commitment (Lambert, Minor, Wells, & Hogan, 2016). Similarly, when people work in a supportive environment, it can prevent and decrease employee turnover (Feeley, Moon, Kozey, & Lowe, 2010). Social support in the workplace helps reduce stress among employees, which positively influences levels of job satisfaction and helps prevent burnout (Singh, Singh, & Singhi, 2015; Snyder, 2009).

7.3 Mentorship

Mentoring is an oft-studied organizational concept that is practiced in prevalence across industries. In fact, mentoring has become somewhat of a buzzword in both empirical and mass media contexts and has been studied both extensively and empirically in U.S. workplaces. While often viewed as a way to promote advancement among disadvantaged groups such as women and/or BIPOC and minorities who lack access to informal and interpersonal career development resources, mentoring programs have been implemented at organizations throughout the United States.

A clear definition of mentoring is a controversial subject, but in essence, mentoring relationships consist between one seasoned or more senior member of an organization and one more junior member of the organization. Mentoring is unique because it can create an unequal and vulnerable relationship. A mentor is also defined as a person who serves as a guide or sponsor to the development of another who has a different rank (Sands, Parson, & Duane, 1991). The most traditional, common, and concise definition of workplace mentoring is a relationship between two people (dyadic) where the senior employee takes the junior employee under his or her wing to share knowledge and provide guidance (Allen, Finkelstein, & Poteet, 2009).

Mentoring has roots in Greek mythology and later evolved into the modern day understanding of a protege, similar to apprenticeship. Mentoring used to be most prevalent among men, where midlife men served as transitional figures for younger men during their early adulthood (Levinson, Darrow, Klein, Levinson, & McKeen, 1978). Mentoring has since evolved into a mentor/mentee relationship and away from the protege conceptualization and practice. Mentees are guided and advised by mentors, who are senior-level role models who provide career guidance, coaching, and support through an ongoing relationship (Darling, 1985; Prehm and Iscson, 1985). Mentees should take an active role in the formation and development of mentoring relationships. Good mentors should be "sincere in their dealings with mentees, be able to listen actively and understand mentees' needs, and have a well-established position within the [organization]." (Sambunjak, Straus, & Marusic, p. 79, 2009).

Mentoring is inherently supportive and provides two forms of support: career-related and psychosocial support. Career-related support focuses on the success, development, and advancement of the mentee. This can include helping the mentee gain exposure and visibility; coaching; providing protection and sponsorship; and providing opportunities for assignments. Psychosocial support, on the other hand, focuses more on the identity of the mentee and building his or her effectiveness as a professional. This includes activities such as helping them make friends, building acceptance and confirmation, and serving as a role model. The psychosocial elements of mentoring have been shown to be positively associated with the outcomes of the mentoring relationship (Allen et al., 2009).

Mentoring is fundamentally different from other types of workplace relationships. First, mentoring is a dyadic relationship between people of two different experience levels. Second, it is both a mutually beneficial relationship but also asymmetrical because the focus of the relationship is on the development of the mentee, despite the benefits that both parties can gain. Finally, mentoring is a fluid and dynamic relationship that changes over time. It differs from supervisor–subordinate relationships, for instance, because the mentor and the mentee are not required to work together, there is no requirement for reward power to be present, and the mentor can be several levels higher than the mentee – or only one level higher (Allen et al., 2009).

Jacobi (1991) explicated five elements that are present in the mentoring relationship: 1) focus on achievement or acquisition of knowledge; 2) emotional and psychological support, direct assistance with career and professional development, and role modeling; 3) reciprocity where both mentor and mentee derive benefits; 4) personal in nature involving direct interaction; and 5) emphasizes the mentor's greater experience, influence, and achievement within a particular organization. This demonstrates the complexity of mentoring and reinforces the notion that leaders are essential to the process of developing supportive and collaborative cultures (Edge, 2014).

Mentoring offers benefits for the mentor, the mentee, and the organization, making it well suited for a multigenerational workforce where people with various levels of experience and knowledge are working together. Mentors, for example, experience enhanced career success, which is great for members for Gen X, career revitalization which can be important for Baby Boomers, and personal growth and satisfaction. Conversely, mentees experience higher compensation and faster salary growth, more promotions and higher expectations for advancement, more job and career satisfaction, and greater commitment to the organization. All of these benefits are very important to Millennial employees, making them well positioned to receive mentoring from a Gen X or Boomer employee. Finally, organizations can experience benefits such as enhanced recruitment and retention efforts and increases in employee socialization and organizational learning (Allen et al., 2009).

As the multigenerational workforce became the norm and organizations had three, sometimes four generations present, a new type of mentoring emerged: reverse mentoring. Reverse mentoring flips traditional mentoring upside down and has a younger, junior employee serving as the mentor to an older, senior colleague (Murphy, 2012). The concept was first introduced by Jack Welsh, CEO of General Electric, but is now considered a best practice at other large organizations including Dell, Estee Lauder, Procter & Gamble, and Time Warner (Greengard, 2002; Harvey & Buckley, 2002; Hewlett, Sherbin, & Sumberg, 2009). One of the main benefits of reverse mentoring is the ability for others to learn from the digital wisdom and savviness of younger generations. Additionally, though, reverse mentoring helps prepare younger employees for leadership roles, fosters diversity efforts, enhances and promotes intergenerational working relationships, and promotes innovation (Murphy, 2012). Furthermore, reverse mentoring, much like traditional mentoring, is a cost-effective professional development strategy for organizations to implement.

7.4 Previous Research and Theory

Mentoring and communication support are frequently studied together because they are antecedents to so many other workplace variables as mentioned throughout this chapter. The mentoring and communication support scale, used in our study, is one of the most commonly applied scales for measurement empirically. The scale was originally developed to study communication within academic organizations but has since been applied to the nonacademic workforce. The scale is useful because it recognizes that mentoring is part of the larger concept of workplace support, while also maintaining the unique attributes of mentoring. The mentoring and communication support scale helps to determine mentoring as well as other supportive behaviors dimensionally (Hill, Bahniuk, Dobos, & Rouner, 1989).

Empirical research utilizing the mentoring and communication support scale has helped researchers to better understand the positive attributes of a supportive workplace that includes mentoring. For instance, Harris, Winskowski, and Engdahl (2007) demonstrated that social support accounted for almost 17% of the variance in job satisfaction, with career mentoring being one of the most predictive factors of job satisfaction. A case study using two large chemical organizations demonstrated that mentoring as part of a larger organizational communication support program helps minority employees climb the corporate ladder, as well as gain friendships, feel supported, and maintain a positive attitude in the workplace (Kogler Hill & Gant, 2000). A management study examined the mutuality present in mentoring and supportive communication finding that when mentees are open to coaching and put forth effort in accomplishing work, mentors' perceptions of relationship effectiveness and trust are positively influenced. Furthermore, when mentors engage in supportive behaviors to meet the expectations of mentees, mentees form higher perceptions of effectiveness and trust for their mentor (Young & Perrewe, 2000). Collectively, this research helps showcase the positive outcomes and importance of mentoring and supportive communication in the workplace, which is further demonstrated through our data in what follows.

7.5 Mentoring and Communication Support and Generational Differences: Our Data

The Mentoring and Communication Support Scale (Hill et al., 1989) indexed participants' experiences of mentoring and communication support in their workplaces. This scale contains four subscales: Career mentoring, coaching, collegial – social and collegial – task. Means across the overall scale and subscales were well over the midpoint, ranging from M = 3.32 for coaching to 4.01 for collegial-task. The coaching subscale was the only scale or subscale to fall just short of the widely recognized .70 criteria for assessing a scale's reliability, so one should be cautious about generalizing findings on that subscale. An examination of differences in this scale and subdimensions by generation are presented below in Table 7.1.

There are no differences between generations and the main scale, as well as three of the four subscales. There is a significant difference in scores on Collegial – Task because Baby Boomers report significantly greater scores than both Gen X and Millennials.

The findings of our generational study are consistent with previous research, which indicates that there is not a generational difference when it comes to mentoring. This helps to underscore the importance of organizational mentoring, though, showcasing that everyone in an

Table 7.1 Generational Differences in the Mentoring and Communication Social Support Scale and Subscales

Measure	*Baby Boomers* M(SD)	*Gen X* M(SD)	*Millennials* M(SD)	F *(2, 1147)*	*eta²*
Mentoring & Communication Social Support	3.69(.71)	3.64(.75)	3.67(.67)	.194	.00
Career Mentoring	3.41(.92)	3.51(.97)	3.60(.87)	2.432	.00
Coaching	3.20(1.03)	3.35(1.01)	3.31(.92)	.850	.00
Collegial – Social	3.77(.87)	3.61(95)	3.69(.86)	1.138	.00
Collegial – Task	**4.27(.80)**[a]	4.03(.89)[b]	3.97(.78)[b]	5.618*	.01

$^{**}p < .001$, $^{*}p < .05$.
[a,b]Significant differences between groups as determined by Tukey HSD post hocs. Scores that significantly differ from the other two scores are bolded.

organization can benefit from the process, regardless of whether they are the mentor or the mentee. Furthermore, with it comes to Baby Boomers, our data indicates that through mentoring, their feelings of social collegiality can improve, which is positive information that organizations should use as additional consideration for implementing mentorship initiatives. Finally, these findings demonstrate that mentoring represents one successful and positive prong in a larger program of supportive workplace programs and practices.

7.6 Best Practices for Creating a Supportive Multigenerational Workplace

The goal of a supportive workplace is to have happy, satisfied, and supported employees because when an organization's employees are happy, the benefits are endless, keeping the organization healthy. The following are some best practices that organizations can implement to improve support, but this is not an exhaustive list, and organizations are encouraged to regularly survey employees and conceptualize new ways to provide support.

Best Practice 1: Provide Enriching Experiences

According to a large workplace survey conducted by MetLife (2019), employees need an ally, and it does not need to be one specific person. Instead, employees need their organization to be their ally, and one way to do this is to provide enrichment opportunities that go beyond employee recognition, and instead, provide support. For instance, workplaces can offer workshops about financial topics or programs to reduce stress (lunchtime meditations, for example); encourage people to use their vacation time; and provide training so that people can do their job better and/or faster to support employees as both organizational assets and humans.

Best Practice 2: Listen

Listening is an important part of supportive communication and is free for organizations to implement. Supportive listening is characterized by focusing attention on the support seeker, expressing involvement, demonstrating understanding, and being both verbally and non-verbally responsive (Mikkola, 2019). In the workplace, anyone can be a supportive listener regardless of age, tenure, position, or power. The more that people engage in supportive listening, the more that people will communicate their needs for support, creating a cyclical process for support in the workplace.

Best Practice 3: Monitor Stress and Manage Uncertainty

There is not one workplace throughout the world where stress and uncertainty is nonexistent, regardless of how supportive the organizational culture is. Stress and uncertainty are workplace norms, but with that, supportive communication can also become a norm, as the concept is intimately linked with stress and uncertainty. One of the easiest ways to create and maintain a supportive work environment is to identify, monitor, and manage stress and uncertainty through communication. Simple tactics such as encouraging employees to discuss stressors and creating a shared need for support (on a project team, for instance) can provide support. Mikkola (2019) suggests reflecting on current communication practices and asking whether or not they promote supportive interactions. For instance, engaging in excessive and/or constant discussion of work and workplace problems can increase stress and lead to defensiveness in communication. Also, the practice of "rush talk" where people are constantly stating how busy and overworked they are promoting a culture of busyness and stress, which is not supportive and can hide communication efforts that seek support. Therefore, engaging in more supportive communication such as asking questions, having honest conversations, and acknowledging stress and uncertainty as hard and discomforting can promote supportive workplaces.

7.7 Conclusion

Similar to Chapter 5, which is about organizational culture, there are certain workplace practices that transcend age and are not influenced by generations. Supportive communication is another one of these items as demonstrated by our data. When a supportive workplace is present, generational effects are minimized, and people coexist with greater ease to help reach organizational goals. Therefore, promoting positive communication in the workplace can provide great benefits and help connect a multigenerational workforce.

References

Allen, T., Finkelstein, L., & Poteet, M. (2009). *Designing workplace mentoring programs: An evidence-based approach*. Wiley-Blackwell.

Bodie, G., & Burleson, B. (2008). Explaining variations in the effects of supportive messages: A dual-process framework. In C. Beck (Ed.), *Communication yearbook* (vol. *32*, pp. 354–398). Routledge.

Brashers, D. (2001). Communication and uncertainty management. *Journal of Communication*, *15*, 477–497.

Burleson, B. (2003). Emotional support skills. In J. O. Green and B. Burleson (Eds.), *Handbook of communication and social interaction skills* (pp. 551–594). Lawrence Erlbaum.

Burleson, B. (2008). What counts as effective emotional support? Explorations of individual and situational differences. In M. T. Motley (Ed.), *Studies in applied interpersonal communication* (pp. 207–227). Sage.

Burleson, B. (2009). Understanding the outcomes of supportive communication: A dual-process approach. *Journal of Social and Personal Relationships*, *26*, 21–38.

Burleson, B., & MacGeorge, E. (2002). Supportive communication. In M. L. Knapp & J. A. Daly (Eds.), *Handbook of interpersonal communication* (3rd ed., pp. 374–424). Sage.

Clark, R., Pierce, A., Finn, K., Hsu, K., Toosley, A., & Williams, L. (1998). The impact of alternative approaches to comforting, closeness of relationship, and gender on multiple measures of effectiveness. *Communication Studies*, *49*, 224–239.

Cohen, S., Mermelstein, R., Karmarck, T., & Hoberman, H. (1985). Measuring the functional components of social support. In I. Sarason & B. Sarason (Eds.), *Social support: Theory, research, and applications* (pp. 73–94). Maritnus Nijhoff.

Cunningham, M., & Barbee, A. (2000). Social support. In C. Hendrick & S. Hendrick (Eds.), *Close relationships: A sourcebook* (pp. 272–285). Sage.

Cutrona, C., Cohen, B., & Igram, S. (1990). Contextual determinants of the perceived helpfulness of helping behaviors. *Journal of Social and Personal Relationships*, *7*, 553–562.

Darling, L. (1985). What to do about toxic mentoring? *Journal of Nursing Administration, May 1985*, *15*, 43–45.

Edge, K. (2014). A review of the empirical generations at work research: Implications for school leaders and future research. *School Leadership & Management*, *34*, 136–155. https://doi.org/10.1080/13632434.2013.869206

Feeley, T., Moon, D., Kozey, R., & Lowe, A. (2010). An erosion model of employee turnover based on network centrality. *Journal of Applied Communication Research*, *38*, 167–188.

Feng, B. (2009). Testing an integrated model of advice-giving in supportive interactions. *Human Communication Research*, *35*, 115–129.

Glynn, L., Christenfeld, N., & Gerin, W. (1999). Gender, social support, and cardiovascular responses to stress. *Psychosomatic Medicine*, *61*, 234–242.

Gottlieb, B. (1994). Social support. In A. Weber & J. Harvey (Eds.), *Perspectives on close relationships* (pp. 307–324). Allyn & Bacon.

Greengard, S. (2002). Moving forward with reverse mentoring. *Workforce, March*, 15.

Grunig, J., & Grunig, L. (2008). Excellence theory in public relations: Past, present, and future. In A. Zerfass, B. van Ruler, & K. Sriramesh (Eds.), *Public relations research: European and international perspectives and innovations* (pp. 327–347). VS Verlag.

Harris, J., Winskowski, A., & Engdahl, B. (2007). Types of workplace social support in the prediction of job satisfaction. *The Career Development Quarterly*, *56*, 150–156.

Harvey, M., & Buckley, M. (2002). Assessing the "conventional wisdoms" of management for the 21st Century organization. *Organizational Dynamics*, *30*, 368–378.

Hewlett, S., Sherbin, L., & Sumberg, K. (2009). Let GenY teach you. Retrieved from https://hbr.org/2009/06/let-gen-y-teach-you-tech

Hill, S. E., Bahniuk, M. H., Dobos, J., & Rouner, D. (1989). Mentoring and other communication support in the academic setting. *Group and Organization Studies*, *14*, 355–368. doi: 10.1177/105960118901400308

Jacobi, M. (1991). Mentoring and undergraduate academic success: A literature review. *Review of Educational Research*, *61*, 505–532.

Jacobson, D. (1986). Types and timing of social support. *Journal of Health and Social Behavior*, *27*, 250–264.

Kogler Hill, S., & Gant, G. (2000). Mentoring by minorities for minorities: The organizational support system. *Review of Business*, *21*, 53–57.

Lambert, E., Minor, K., Wells, J., & Hogan, N. (2016). Social support's relationship to correctional staff job stress, job involvement, job satisfaction, and organizational commitment. *Social Science Journal*, *53*, 22–32.

Levinson, D., Darrow, C., Klein, E., Levinson, M., & McKeen, B. (1978). *Seasons of a man's life*. Knopf.

MacGeorge, E., Feng, B., & Thompson, E. (2008). "Good" and "bad" advice: How to advise more effectively. In M. Motley (Ed.), *Studies in applied interpersonal communication* (pp. 145–164). Sage.

MetLife (2019). Thriving in the new work-life world. MetLife's 17th annual U.S. employee benefit trends study. Retrieved from https://www.metlife.com/content/dam/metlifecom/us/ebts/pdf/MetLife-Employee-Benefit-Trends-Study-2019.pdf

Mikkola, L. (2019). Supportive communication in the workplace. In L. Mikkola & M. Valo (Eds.), *Workplace communication* (pp. 147–162). Taylor & Francis.

Murphy, W. (2012). Reverse mentoring at work: Fostering cross-generational learning and developing Millennial leaders. *Human Resource Management*, *51*, 549–574.

Neff, L., & Karney, B. (2005). Gender differences in social support: A question of skill or responsiveness? *Journal of Personality and Social Psychology*, *7*, 561–570.

Park, K., Wilson, M., & Lee, M. (2004). Effects of social support at work on depression and organizational productivity. *American Journal of Health Behavior*, *28*, 444–455.

Prehm, H., & Iscson, S. (1985). Mentorship: Student and faculty perspectives. *Teacher Education and Special Education: The Journal of the Teacher Education Division of the Council for Exceptional Children*, *8*, 12–16.

Sambunjak, D., Straus, S., & Marusic, A. (2009). A systematic review of qualitative research on the meaning and characteristics of mentoring in academic medicine. *Journal of General Internal Medicine*, *25*, 72–78.

Sands, R., Parson, L., & Duane, J. (1991). Faculty mentoring faculty in a public university. *The Journal of Higher Education*, *62*, 174–193.

Shaver, P. (2008, Spring). Some necessary links between communication studies and social psychology in research on close relationships. *Relationship Research News*, *6*, 1–2.

Singh, A., Singh, A., & Singhi, N. (2015). Organizational role stress and social support as predictors of job satisfaction among managerial personnel. *Journal of Social Service Research*, *40*, 178–188.

Snyder, J. (2009). The role of coworker and supervisor social support in alleviating the experience of burnout for caregivers in the human-services industry. *Southern Communication Journal*, *74*, 373–389.

Uno, D., Uchino, B., & Smith, T. (2002). Relationship quality moderates the effect of social support given by close friends on cardiovascular reactivity in women. *International Journal of Behavioral Medicine*, *9*, 243–262.

Young, A., & Perrewe, P. (2000). What did you expect? An examination of career-related support and social support among mentors and proteges. *Journal of Management*, *26*, 611–632.

8 The Dark Side of Communication at Work: Conflict and Dissent

This chapter will explore destructive workplace communication – specifically, conflict and dissent. The authors will reflect on their findings and connect these findings to previous literature surrounding conflict and dissent. This chapter will address multigenerational workplace preferences, including crisis communication in the work environment.

8.1 Ineffective Communication in the Workplace

In an ideal world, employees would understand supervisor expectations clearly, tasks would be accurately completed, and workplace relationships would be positive and supportive. Unfortunately, our world is not idealistic, and often workplaces struggle to create an environment that is conducive to positive communication. Indeed, communication is crucial to organizational structures, with Odine (2015) even remarking that management can only thrive in the prevalence of communication. A positive and supportive communication climate does not happen naturally and instead must be nurtured and developed. Our organizations are, probably, more likely to engage in and practice ineffective communication.

In 2011, Chandra, Theng, Lwin, and Shou-Boon (2011) identified uncertainty as one of the primary barriers to effective communication in the workplace. The concept of uncertainty tends to lead to equivocality or ambiguity. These barriers are more conceptual, relying on the message and the sender. Other barriers exist, including cultural and language differences; literal physical barriers that impact noise; psychological barriers like self-esteem, jargon, and language; differing expectations; or even differing abilities. This list is nowhere near exhaustive. Yet, despite all of these challenges, for organizations to truly develop an environment where employees want to work and to avoid significant turnover (Nwagbara, Oruh, Ugorji, & Ennsra, 2013), communication should be a key component of an organization's mission. Ineffective communication can lead to decreased productivity, satisfaction, and lower retention rates (Salahuddin, 2010).

In some ways, ineffective communication may also boil down to employee expectations. Millennials, for one, prefer organizations that have a reduced hierarchical structure (Barnes, 2009). For this group, then, communication in the workplace that is ineffective would seem to reinforce "red tape" or decrease transparency. While this is not the only communication difference between generations, it is prevalent.

A deteriorating focus on communication can have significant ramifications beyond those traditionally reserved for job functionality. Specifically, ineffective communication can become inherently negative. Workplace bullying, for one, can occur because of diminished communication channels (Venkataramani, Labianca, & Grosser, 2013). Gossip can also occur when communication expectations have not been defined clearly (Ye, Zhu, Deng, & Mu, 2019). These consequences highlight the dark side of workplace communication, but there are other manifestations of negative communication in the workplace as well.

8.2 Manifestations of Negative Communication

Just because communication is ineffective does not mean it is inherently negative. In fact, ambiguity in some cases does not manifest itself as either a positive or negative outcome. Instead, it could occur because of poor planning or basic misunderstanding. Yet, there are instances where communication in the workplace becomes a negative event with negative results. Negative messages cause unpleasant reactions even as intense as sickness, absences, reduced motivation, and productivity declines leading to financial consequences (Kline & Lewis, 2019). Generally, negative communication leads to negative relationships in the workplace (Keashly & Jagatic, 2003). Workers, struggling to control a negative communication environment, will see increased stress, emotional load, turnover, and cynicism toward the organization and their life in general (Fritz & Omdahl, 2006). Negative communicative behaviors can be long-lasting affairs or isolated incidents. No matter the duration, negative behaviors can lead to long-term emotional strain.

Negative communicative behaviors and problematic relationships can manifest through different relational categorizations. Tuikka (2020) believes the most problematic negative relationships in the workplace are those that are uncivil, aggressive, (sexually) harassing, unwanted, or are defined by conduct that is unprofessional. Organizations should clearly identify their expectations for handling unethical, uncivil, and unprofessional communication (Fritz, 2019).

Uncivil Relationships

Incivility is pervasive in organizations (Cortina & Magley, 2009). While the definition of incivility, especially in the workplace, is somewhat ambiguous (Andersson & Pearson, 1999), it is important to note that there is some form of deviance or violated norms for respect associated with incivility. Incivility may be spurred on by many factors, but perceived intent to harm may be one of the most influential factors that contributes to employee mistreatment. No matter what the reasons, organizations should determine that civility will be a central tenet of their culture.

Harassment

Harassment, especially sexual harassment, is widespread (Hardies, 2019). McDonald (2012) defines sexual harassment as behaviors toward targets and can include unwanted sexual comments, propositions or requests, gestures, or even actions and assault. Like other negative communication, sexual harassment can lead to mental and physical health challenges and decreased job performance (Willness, Steel, & Lee, 2007).

Incivility and harassment can generally be classified as behavior that is unprofessional. Unprofessional behavior, and negative communication, fuel dysfunctional organizational culture. And, especially because there are now additional outlets, like social media platforms, for these dark side behaviors to occur, organizations would be wise to address their policies and procedures related to these actions.

Unfortunately, organizations that struggle to handle this dark side of workplace communication may struggle to adapt and evolve in an environment increasingly concerned with positive and supportive climates. As virtual work has become more prominent, tasks that at one time were completed face to face are now completed in computer-mediated environments (Vrankes, Bailien, Vandebosch, Erreygers, & De Witte, 2017). Organizations, therefore need to focus not only on negative communication in physical contexts but also on negative or damaging communication in virtual environments. All of this to say: ignoring the dark side of workplace communication is not a luxury the 21st century organization can afford. Negative communication has far-reaching effects. For our purposes, though, two of the most pressing consequences of negative communication include conflict and dissent.

8.3 Conflict

8.3.1 Previous Research and Theory

Conflict in the workplace is inevitable and occurs in all organizations (Tuikka, 2020). However, the actual manifestation of conflict, and for our purposes the analysis of conflict in the workplace, can be an area of disagreement (Chaudry & Asif, 2015). No matter how conflict in organizations comes to fruition, it has been a significant area of study for researchers (Litterer, 1966).

The definition of conflict is debated. The range of different definitions conceptualizes conflict as a reflection of interpersonal hostility (Barki & Hartwick, 2001), a phenomenon that includes emotions, perceptions, and behaviors (Pondy, 1969), or even disagreement of how to achieve certain goals Jehn, 1997). Chaudry and Asif (2015) believe the common theme of these varying definitions revolves around identifying what triggers and prolongs conflict. They, then, conceptualize conflict as a "cohesive framework of behavior and perception of organizational members, which is triggered by the feelings of being deprived with an awareness of incompatibility with others" (Chaudry & Asif, 2015, p. 239). This definition is helpful because it brings to light a condition where conflict arises because one does not get what they want or finds a relationship incompatible.

Like the actual definition of conflict itself, expert opinions differ on the benefits of conflict. Some view conflict as a functional dynamic (Chen, 2006; Harolds & Woods, 2006; Jehn & Bendersky, 2003), while others view conflict as damaging to the organizational structure (De Dreu, 2008; Litterer, 1966). Recognizing that individual situations and organizations probably have different outcomes related to their conflict, it is worth noting that the results of conflict probably depend more on organizational culture and personal relationships within the organization and are not necessarily inherently positive or negative (Sharma & Singh, 2019). Ultimately, individuals in organizations will handle conflict differently.

From a generational perspective, Dencker, Joshi, and Martocchio, 2008 believe that an accurate understanding of generational differences in the workplace that help mitigate conflict and enhance conflict resolution. Conflict can occur in workplaces between generations for a variety of reasons. Urick, Hollensbe, and Masterson (2012) identify three different forms of intergenerational conflict: value based, behavior based, and identity based. Intergenerational work conflict can occur because of several different factors but miscommunication, work-life balance, technology-use differences (Carver & Candela, 2008), and issues with older/younger supervisor dyadic relationship challenges (Collins, Hair, & Rocco, 2009). Unfortunately, if not handled properly, generational conflict in the workplace can negatively impact the organization (Sessa, Kabacoff, Deal, & Brown, 2007).

To measure conflict in organizations, Putnam and Wilson (1982) developed the Organizational Communication Conflict Instrument (OCCI). Their instrument, unlike those prior, focuses on concrete communicative behaviors, not just conflict styles. We used this measure to identify generational differences in endorsements of various conflict strategies in the workplace. Their instrument asked participants how often they use a particular strategy. Specifically, their strategies were nonconfrontation strategies (generally avoid disagreements, downplaying controversies, or approaching conflict indirectly); solutions-oriented strategies (using compromise or a search for innovation); and finally control strategies (those behaviors that seek to manage conflict by arguing for particular positions). These three subdimensions help to inform our understanding of how employees engage in actual communication behaviors related to conflict.

8.3.2 Conflict and Generational Differences: Our Data

The Organizational Communication Conflict Instrument (Putnam & Wilson, 1982) was used to index endorsement of three different conflict strategies: Nonconfrontation, solutions-oriented, and control strategies. The scale and its three subscales all demonstrated reliability coefficients > .82. Greater numbers on this (7-point) scale indicate greater agreement that individuals adopt each of the three strategies, where Solutions-oriented strategies has the greatest average, $M = 4.77$, $SD = .94$), followed by nonconfrontation strategies ($M = 3.97$, $SD = 1.13$) and lastly control strategies ($M = 3.25$, $SD = 1.13$). An analysis of differences across the three generations is presented in Table 8.1.

Results show overall similar endorsement of each of the three strategies, with significant differences emerging in nonconfrontation strategies. Here, we see Millennials endorse this strategy to a greater extent than both Gen X and Baby Boomers.

8.4 Dissent

8.4.1 Previous Research and Theory

Conflict can, potentially, lead to organizational dissent. Dissent, according to Kassing (1998), is “how employees verbally express their

Table 8.1 Generational Differences in the Organizational Communication Conflict Instrument and Subscales

Measure	*Baby Boomers* M(SD)	*Gen X* M(SD)	*Millennials* M(SD)	F *(2, 1147)*	*eta*^{2}
Organizational Com. Conflict Instrument	3.93(.65)[a]	3.95(.63)[a]	**4.15(.63)**[b]	12.432**	.02
Non-Confrontation Strategies	3.64(1.15)[a]	3.74(1.22)[a]	**4.07(1.09)**[b]	12.454**	.02
Solutions-Oriented Strategies	4.79(.98)	4.66(1.02)	4.81(.91)	2.262	.00
Control Strategies	3.10(1.11)	3.20(1.17)	3.28(1.13)	1.201	.00

**$p < .001$, *$p < .05$.
[a,b]Significant differences between groups as determined by Tukey HSD post hocs. Scores that significantly differ from the other two scores are bolded.

contradictory opinions and disagreements about organizational phenomena" (p. 183). Dissent, then, manifests in the sharing of employee opinions about the organization. As one can imagine, a negative communication climate can lead to high levels of organizational dissent. Theoretically dissent is expressed when employees, specifically, share "contradictory opinions about organizational practices, policies, and operations" (Kassing, 1998, p. 183).

Generational differences surrounding dissent and the expression of dissent in organizations tend to vary. For instance, Shakil and Siddiqui (2020) report that when millennials are unable to express their dissent, the situation becomes more detrimental, especially to their sense of commitment to the organization. Boomers, on the other hand, tend to be a little more positive related to their overall work values and report that they view work as an extension of their self-interests and as a place where they can experience gratification and growth (Zemke, Raines, & Filipczak, 1999). Gen Xers tend to represent the opposite end of the spectrum. Members of Generation X spent their formative years in organizations where downsizing and outsourcing were common practices (Jones & Murray, 2019). Ironically, Millennials tend to have higher levels of overall company satisfaction compared to Xers (Kowske, Rasch, & Wiley, 2010) yet Millennials are more inclined to communicate their dissenting thoughts to coworkers and supervisors (Myers & Sadaghiani, 2010). These realities create a difficult communication environment for employees and supervisors to navigate.

One way organizations can create a safe environment for employees to share organizational feedback is by recognizing how dissent is communicated in their organization. To measure organizational dissent, Kassing (2000), developed an 18-item Organizational Dissent Scale, which indexes overall dissent toward the organization along two different sub-dimensions: articulated and latent.

Articulated Dissent

Employees who believe their dissent will be perceived more favorably tend to practice articulated dissent. In instances where articulated dissent is expressed, employees assume their thoughts will be perceived as helpful or constructive and that the hearers will not retaliate against the employee who voices the concern (Kassing, 1997). Employees, too, who express their dissent in this way tend to communicate to those who they believe can actually make some change.

Latent Dissent

Unlike articulated dissent, latent dissent, originally conceptualized as antagonistic dissent, tends to be more adversarial. Those who express latent dissent tend to believe they are still safe from retaliation because they possess an organizational leverage (Kassing, 1997). The term latent dissent also expresses a nonobservable concept meaning, dissent readily exists but may not be easy to measure or observe, and as the dissent grows, the possibility for observation tends to increase.

Workplace communication, specifically negative workplace communication, has substantial implications on day-to-day operations and productivity. Even in "normal" workplace scenarios, it can be hard to manage so many personalities, relationships, and generational differences. Yet, in crisis situations, organizations should work even harder to create environments that are transparent and effective. To combat negative communication in crisis that leads to conflict and dissent, organizations should develop a strategy that clearly identifies a crisis communication strategy and different ideas for communication in a multigenerational workplace during crisis situations.

Table 8.2 Generational Differences in the Organizational Dissent Scale and Subscales

Measure	*Baby Boomers* M(SD)	*Gen X* M(SD)	*Millennials* M(SD)	F (*2, 1147*)	*eta*^{2}
Organizational Dissent Scale	2.94(.58)	3.03(.68)	3.02(.58)	.893	.00
Articulated Dissent	**2.37(.74)**[a]	**2.66(.86)**[b]	**2.86(.82)**[c]	17.809**	.03
Latent Dissent	3.50(.84)[a]	3.40(.911)[a]	**3.18(.70)**[b]	13.619**	.02

$^{**}p < .001$, $^{*}p < .05$.
[a,b]Significant differences between groups as determined by Tukey HSD post hocs. Scores that significantly differ from the other two scores are bolded.

8.4.2 Dissent and Generational Differences: Our Data

Kassing's (2000) Organizational Dissent Scale demonstrated reliability (Cronbach's alpha = .824). Across all participants, scores were higher for Latent Dissent (M = 3.25, SD = .77) than for Articulated Dissent (M = 2.78, SD = .83), on a 5-point scale. Analyses of differences across generations are presented in Table 8.2.

Interesting, significant differences are found across articulated and latent dissent among the generations. Notably, Millennials stick out as demonstrating the *greatest* articulated dissent and the *least* latent dissent. This finding reflects previous research that Millennials are more likely to be more vocal about their dissent in the workplace. The opposite pattern is true for Baby Boomers, who show the least articulated dissent and the greatest latent dissent (although they do not differ from nearly as low Gen X in latent dissent.

8.5 Multigenerational Crisis Response

In organizations, as employers struggle to overcome negative communication, conflict, and dissent, a firm grasp of crisis management is also important for the modern manager. Crisis researchers have primarily focused on external messaging in crisis situations (Frandsen & Johansen, 2011), but internal stakeholders need crisis leadership as well. And, in our modern organizations where incivility, harassment, bullying, and other dark side interactions are common, a crisis approach can be helpful. Crisis situations, referred by Mitroff (2005) as major acts of betrayal, should be approached with wisdom and intentionally. In turn, employers should strive for an element of transparency as they deal with internal crisis situations. Transparency, according to Schnackenberg and Tomlinson (2016), is the "perceived quality of intentionally shared information from a sender" (p. 1788). A transparency initiative will speak volumes to Millennials and Generation Z. As Stewart, Oliver, Cravens, and Oishi (2017) point out, increased transparency can help increase an employee's drive and increase operational efficiency.

External relationships tend to drive crisis communication; however, as a potential categorization of negative or dark-side internal communication, crises can be approached similarly to other employee communication. Strategic communication between managers and internal stakeholders should, generally, promote commitment and organizational belonging (Welch, 2012).

Historically, research has conceptualized crisis communication based on which channels are utilized and/or how frequently messaging occurs; however, this approach may fall short of what organizations actually need to know and share about their internal crisis strategy (Ruck & Welch, 2012). While it is important to consider both frequency and channel in crisis situations, the actual content and the impending dialogue brought about by messages should be a focal point of employers. However, channels used to communicate in crisis situations should be adaptable and, from a generational perspective, managers must be more adept today at using technology to respond to crisis situations in organizations (Vielhaber & Waltman, 2008) because new or efficient technologies can be used to communicate with different generational groups, and younger generations prefer high-tech channels.

8.6 Best Practices for Handling Conflict, Dissent, and Crisis in the Modern Multigenerational Organization

The consideration of employers and managers, when dealing with all that goes into the dark side of corporate communication, is a topic not for the faint of heart. The institutional ramifications of negative communication, including but limited to conflict and dissent, as well as potential crisis situations should be approached thoughtfully. To offer practical solutions or guidance for dealing with negative communication in the workplace, the following best practices are proposed.

Best Practice 1: Establish a Culture of Advocacy and when Necessary, Appropriate Policies

Organizations must work to establish an organizational culture that emphasizes equitability, advocacy, and inclusion. Along those lines, establishing a culture where people understand how to be in relationships with each other can create a tone of positive normalcy. Meaning, positive interaction can become the norm not the exception. However, to do this, it may be appropriate to establish policies for behavior that is unacceptable. When you communicate clearly to your people that bullying, harassment, and other manifestations of negative communication will not be tolerated, it can shed light on your organizational values. In addition, clarifying for your people how to deal with conflict and dissent, giving appropriate time and context for disagreement, can also be helpful. In sum, normalize the process for healthy conflict and dissent and formalize a policy against unhealthy negative communication and behavior.

Best Practice 2: Be Transparent (When You Can)

From a generational perspective, transparency can be a nonnegotiable desire for Millennials and members of Generation Z. Generation X and Baby Boomers have less demanding expectations for transparency and accountability from their bosses and supervisors, but they still desire open and appropriate communication. Be transparent when you can. Let your people know how you are dealing with internal crises, inform them of your rationale for decisions, and communicate information that they need to know. Obviously, you cannot be transparent about everything, but ask yourself if you are sharing everything you should share.

Best Practice 3: Do not Allow Dissent to Fester

Employees will always have dissenting opinions. However, pretending that employees are on board with every decision and that they do not need a chance to share their perspective can lead to a dangerous outcome. Dissent can be helpful because it provides additional action steps and differentiated operations. Do not be afraid of dissent or conflict, but make sure you have identified your own personal style for dealing with both then provide opportunities for healthy feedback. Ignoring dissenting opinions or failing to provide opportunities for employees to share differing thoughts in a safe space can allow dissent to fester and become a negative cultural distinction. Generally, people appreciate the opportunity to share their thoughts, and younger generations in particular – those more willing to be explicit about their opinions – will generally take you up on opportunities you provide. As you think about a platform for employee sharing, determine if the feedback should be anonymous, think about follow-up mechanisms, and provide clear communication about how feedback will be received. Clearly communicate a cycle for feedback and follow-up.

8.7 Conclusion

Our data show an interesting parallel to previous research. For one, Millennials are not afraid to voice their concerns and opinions, as they relate to the workplace. This new cultural dimension, one of unabashed vocal feedback and opinion sharing, is relatively new and was not a sustained characteristic of Gen X or Boomer employees. In addition, we see, ironically, Millennials appreciate nonconfrontational strategies when dealing with workplace conflict. This particular group, those Generation Y (i.e. Millennials) workers, again show a new and unique way of dealing with issues and challenges in the modern work environment.

References

Andersson, L. M., & Pearson, C. M. (1999). Tit for tat? The spiraling effect of incivility in the workplace. *Academy of Management Review*, *24*, 452–471.

Barki, H., & Hartwick, J. (2001). Interpersonal conflict and its management in information system development. *MIS Quarterly*, *25*, 195–228.

Barnes, G. (2009). Guess who's coming to work: Generation Y. Are you ready for them? *Public Library Quarterly*, *28*, 60.

Carver, L., & Candela, L. (2008). Attaining organizational commitment across different generations of nurses. *Journal of Nursing Management*, *16*, 984–991. doi: 10.1111/j.1365-2834.2008.00911.x

Chandra, S., Theng, Y., Lwin, M. O., & Shou-Boon, S. (2011, May 26–30). *Exploring trust to reduce communication barriers in virtual world collaborations* [paper]. 60th Annual International Communication Association (ICA) Conference, Boston. https://www.researchgate.net/profile/May_Lwin/publication/267381074_Exploring_Trust_to_Reduce_Communication_Barriers_in_Virtual_World_Collaborations/links/5474abb10cf2778985abf047.pdf

Chaudry, A. M., & Asif, R. (2015). Organizational conflict and conflict management: A synthesis of literature. *Journal of Business and Management Research*, *9*, 238–244.

Chen, M. H. (2006). Understanding the benefits and detriments of conflict on team creativity process. *Creativity and Innovation Management*, *15*, 105–116.

Collins, M., Hair, J., & Rocco, T. (2009). The older-worker-younger-supervisor dyad: A test of the reverse Pygmalion effect. *Human Resource Development Quarterly*, *20*, 21–41. doi: 10.1002/hrdq.20006

Cortina, L. M., & Magley, V. J. (2009). Patterns and profiles of response to incivility in the workplace. *Journal of Occupational Health Psychology*, *14*, 272–288.

De Dreu, C. K. (2008). The virtue and vice of workplace conflict: Food for (pessimistic) thought. *Journal of Organizational Behavior*, *29*, 5–18.

Dencker, J. C., Joshi, A., & Martocchio, J. J. (2008). Towards a theoretical framework linking generational memories to workplace attitudes and behaviors. *Human Resource Management Review*, *18*, 180–187.

Frandsen, F., & Johansen, W. (2011). The study of internal crisis communication: Towards an integrative framework. *Corporate Communications: An International Journal*, *16*, 347–361. https://doi.org/10.1108/13563281111186977

Fritz, J. M. H. (2019). Communicating ethics and bullying. In R. West & C. S. Beck (Eds.), *The Routledge handbook of communication and bullying* (pp. 22–29). Routledge.

Fritz, J. M. H., & Omdahl, B. L. (2006). *Problematic relationships in the workplace*. Peter Lang.

Hardies, K. (2019). Personality, social norms, and sexual harassment in the workplace. *Personality and Individual Differences*, *151*, 1–5. https://doi.org/10.1016/j.paid.2019.07.006

Harolds, J., & Wood, B. P. (2006). Conflict management and resolution. *Journal of the American College of Radiology*, *3*, 200–206.

Jehn, K. A. (1997). A qualitative analysis of conflict types and dimensions in organizational groups. *Administrative Science Quarterly*, *42*, 530–557.

Jehn, K. A., & Bendersky, C. (2003). Intragroup conflict in organizations: A contingency perspective on the conflict outcome relationship. *Research in Organizational Behavior*, *25*, 187–242.

Jones, J. S., & Murray, S. R. (2019). The effect of generational differences on work values and attitudes. *International Journal of Research in Business and Management*, *1*, 25–35.

Kassing, J. W. (1997). Articulating, agonizing, and displacing: A model of employee dissent. *Communication Studies*, *48*, 311–331.

Kassing, J. W. (1998). Development and validation of the organizational dissent scale. *Management Communication Quarterly*, *12*, 183–229.

Kassing, J. W. (2000). Investigating the relationship between superior-subordinate relationship quality and employee dissent. *Communication Research Reports*, *17*, 58–70.

Keashly, L., & Jagatic, K. (2003). By any other name: American perspectives on workplace bullying. In E. Einarsen, H. Hoel, D. Zapf, & Cooper, C., Eds., *Bullying and emotional abuse in the workplace: International perspectives in research and practice* (pp. 31–36). Taylor & Francis.

Kline, R., & Lewis, D. (2019). The price of fear: Estimating the financial cost of bullying and harassment to the NHS in England. *Public Money and Management*, *39*, 166–174.

Kowske, B., Rasch, R., & Wiley, J. (2010). Millennials' (lack of) attitude problem: An empirical examination of generational effects on work attitudes. *Journal of Business & Psychology*, *25*, 265–279

Litterer, J. A. (1966). Conflict in organization: A reexamination. *Academy of Management Journal*, *9*, 178–186.

McDonald, P. (2012). Workplace sexual harassment 30 years on: A review of the literature. *International Journal of Management Reviews*, *14*, 1–17.

Mitroff, I.I. (2005). *Why some companies emerge stronger and better from a crisis*. Amacom.

Myers, K., & Sadaghiani, K. (2010). Millennials in the workplace: A communication perspective on Millennials' organizational relationships and performance. *Journal of Business & Psychology*, *25*, 225–238.

Nwagbara, U., Oruh, E. S., Ugorji, C., & Ennsra, M. (2013). The impact of effective communication on employee turnover intension at First Bank of Nigeria, *4*, 13–21.

Odine, M. (2015). Communication problems in management. *Journal of Emerging Issues in Economics, Finance and Banking*, *4*, 1615–1630.

Pondy, L. R. (1969). Varieties of organizational conflict. *Administrative Science Quarterly*, *14*, 499–505.

Putnam, L. L., & Wilson, C. E. (1982). Communicative strategies in organizational conflicts: Reliability and validity of a measurement scale. *Communication Yearbook*, *6*, 629–652.

Ruck, K., & Welch, M. (2012). Valuing internal communication: Management and employee perspectives. *Public Relations Review*, *38*, 294–302. doi:10.1016/j.pubrev.2011.12.016

Salahuddin, M. M. (2010). Generational differences impact on leadership style and organizational success. *Journal of Diversity Management*, *5*, 1–6. https://doi.org/10.19030/jdm.v5i2.805

Schnackenberg, A. K., & Tomlinson, E. C. (2016). Organizational transparency: A new perspective on managing trust in organization-stakeholder relationships. *Journal of Management*, *42*(7), 1784–1810. doi: 10.1177/0149206314525202

Sessa, V., Kabacoff, R., Deal, J., & Brown, H. (2007). Research tools for the psychologist-manager: Generational differences in leader values and leadership behaviors. *Psychologist-Manager Journal*, *10*, 47–74. doi: 10.1080/10887.150701205543

Shakil, B., & Siddiqui, D. A. (2020). Factors affecting millennials employees' dissent and its subsequent impact on their commitment. Retrieved from https://ssrn.com/abstract=3683231

Sharma, S., & Singh, K. (2019). Positive organizational culture: Conceptualizing managerial role in interpersonal conflict. *European Journal of Business & Social Sciences*, *7*, 1508–1518.

Stewart, J. S., Oliver, E. G., Cravens, K. S., & Oishi, S. (2017). Managing millennials: Embracing generational differences. *Business Horizons*, *60*, 45–54. https://doi.org/10.1016/j.bushor.2016.08.011

Tuikka, S. (2020). Negative relationships in the workplace. In L. Mikkola & M. Valo (Eds.), *Workplace communication*. Routledge.

Urick, M. J., Hollensbe, E. C., & Masterson, S. S. (2012). Understanding and managing generational tensions. Presented at Academy of Management Annual Meeting, Boston, MA.

Venkataramani, V., Labianca, G. J., & Grosser, T. (2013). Positive and negative workplace relationships, social satisfaction, and organizational attachment. *Journal of Applied Psychology*, *98*(6), 1028–1039. https://doi.org/10.1037/a0034090

Vielhaber, M. E., & Waltman, J. L. (2008). Changing uses of technology: Crisis communication responses in a faculty strike. *The Journal of Business Communication (1973)*, *45*(3), 308–330.

Vrankes, I., Bailien, E., Vandebosch, H., Erreygers, S., & De Witte, H. (2017). The dark side of working online: Towards a definition and an emotion reaction model of workplace cyberbullying. *Computer in Human Behavior*, *69*, 324–334. https://doi.org/10.1016/j.chb.2016.12.055

Welch, M. (2012). Appropriateness and acceptability: Employee perspectives of internal communication. *Public Relations Review*, *38*, 246–254. doi:10.1016/j.pubrev.2011.12.017

Willness, C. R., Steel, P., & Lee, K. (2007). A meta-analysis of the antecedents and consequences of workplace sexual harassment. *Personnel Psychology*, *60*, 127–162.

Ye, Y., Zhu, H., Deng, X., & Mu, Z. (2019). Negative workplace gossip and service outcomes: An explanation from social identity theory. *International Journal of Hospitality Management*, *82*, 159–168. https://doi.org/10.1016/j.ijhm.2019.04.020

Zemke, R., Raines, C., & Filipczak, B. (1999). *Generations at work: Managing the class of veterans, boomers, Xers, and nexters in your workplace*. AMACOM.

9 Workplace Satisfaction

This chapter focuses on generational perspectives on workplace satisfaction.

9.1 Communication and Workplace Satisfaction

Extensive research on the relationship between communication satisfaction, first conceptualized by Downs and Hazen (1977), and job satisfaction did not begin until the mid-1970s (Pincus, 1986). Since that time, communication satisfaction has become a stable construct in organizational communication research (Crino & White, 1981). However, more than two decades ago, Pincus (1986) noted the lack of research examining the relationships between communication satisfaction, job satisfaction, and communication channel. Surprisingly, given the rapidly expanding capabilities of communication technologies, research in this area is still developing, and with the changes of 2020, this will remain a hot topic within empirical organizational research and for organizations.

Within the organizational setting, communication satisfaction is defined as "an individual's satisfaction with various aspects of communication in his organization" (Crino & White, 1981, pp. 831–832). Akkirman and Harris (2005) measured the following six factors: employees' relationship with supervisor, communication climate, overall communication satisfaction, horizontal communication, organizational integration, and personal feedback. The researchers discovered that teleworkers experience higher levels of communication satisfaction on all factors compared to traditional office workers. Tsai and Chuang (2009) found that supervisory communication, personal feedback, and communication climate are the greatest contributors to the communication-job performance relationship among employees. Although these two studies investigated the broader organizational communication structures, past research has not significantly accounted for the role of interpersonal communication satisfaction and job satisfaction among employees.

Interpersonal communication satisfaction within the workplace can be defined as "an employee's overall affective reaction to his or her evaluation of interaction patterns with coworkers across situations and levels within an organization" (Park & Raile, 2010, p. 572). Interpersonal relationships play a significant role in the satisfaction of employees as it relates to various constructs including communication satisfaction. Informal co-worker interactions are important for building synergy within the organization (Kurland & Bailey, 1999). In addition, satisfying interpersonal relationships are crucial for effective performance and the spreading of organizational culture (Watson-Manheim & Belanger, 2007). Furthermore, research has found the necessity of physical proximity in the development of co-worker relationships to be the least important factor for communication satisfaction (Sias, Pedersen, Gallagher, & Kopaneva, 2012).

It is challenging to separate communication satisfaction from job satisfaction because the two concepts are so related. However, typically if someone is very dissatisfied with their job, it relates back to a communication issue, indicating that communication satisfaction is an antecedent to job satisfaction. There are exceptions to this, of course, but understanding communication satisfaction independently before exploring job satisfaction can help discern when communication is going well and when communication practices need work so as not to disrupt a person's overall job satisfaction.

9.1.1 Previous Research and Theory

Two theories from previous research can help explain workplace communication satisfaction; employers will find these theories useful, especially when it comes to understanding the preferences of a multigenerational workforce. Understanding these two theories can help prevent miscommunications and communication dissatisfaction with a multigenerational workforce, leading to more effective communication and better outcomes among employees. The two theories are: media richness theory and channel expansion theory, which are commonly studied together.

Media Richness Theory

Media richness theory has emerged as one of the most widely studied and cited frameworks in the body of research on organizational media use (D'Urso & Rains, 2008). The premise of media richness theory is that a communication medium should be consistent with the needs of the message for effective communication (Lengel & Daft, 1988). Richness concerns a medium's capacity to convey various types of information cues in a manner that approximates face-to-face communication (Sheer, 2011). Media richness follows a continuum from high richness (i.e. face-to-face) to low richness (i.e. bulletin boards) for understanding the transmission of messages. The richness of a medium comprises four aspects: (1) the availability of instant feedback, which allows questions to be asked and answered; (2) the use of multiple cues, such as physical presence, vocal inflection, body gestures, words, numbers, and graphic symbols; (3) the use of natural language, which can be used to convey an understanding of a broad set of concepts and ideas; and (4) the personal focus of the medium (Lengel & Daft, 1988). When none or only a few of these attributes are present, a medium is considered "lean" (Sheer, 2011). Therefore, face-to-face is considered the richest medium because it allows for all four aspects important in communication.

The most effective choice of media is one that matches the intended outcome for a message, which indicates whether a rich or lean media should be utilized for message distribution (Easton & Bommelje, 2011). Sheer and Chen (2004) demonstrated that rich media have greater personal information-carrying capacities than lean media as analogous to communication immediacy. Additional research has favored face-to-face communication, pointing out that humans are most accustomed to "natural" characteristics only present in face-to-face communication (Kock, 2004) and that face-to-face interactions hold social advantages not present in other forms of media communications (Green et al., 2005).

According to the theory, messages should be communicated on channels with appropriate richness capabilities. When information is communicated using an inappropriate channel, the information is likely to be misinterpreted or seen as ineffective with regard to the intended purpose (Carlson & Zmud, 1999). Additionally, when a message and medium mismatch occurs, communication parties have to engage in compensating communication activities, which takes additional time and resources (Hollingshead, McGrath, & O'Connor, 1993). Media richness research has spent a considerable amount of time identifying the limitations of one channel versus others. For example, Jacobsen (1999) argued that new media, such as instant messaging and online communication, are limited in conveying the same amount of information as a face-to-face conversation. Specifically, when two communicators are not in the same place, physical contact and other nonverbal cues such as olfactory cues become impossible (Kock, 2004). Henderson and Gilding (2004) illustrated that communicating via lean mediums could affect the effectiveness and amount of self-disclosure, thereby influencing reciprocity and trust. While this study was conducted within an interpersonal context, these findings suggest organizational implications. Sheer's (2011) study discovered that a popular reason for the abundance of instant message use between friends is the ability to control information and self-presentation, which could also be applied to co-worker relationships. Additionally, D'Urso and Rains (2008) found support indicating that richness is based on perception and that richness may be shaped by interpersonal factors, such as one's relevant experiences, which is another finding that may lend insight into communication channel satisfaction between co-workers.

As new technologies increasingly become integrated into organizations, the channels of communication available to employees continue to expand. Media channels vary greatly in their richness (Lengel & Daft, 1988). Flyers and bulletins are considered the leanest form of communication, as these are limited in their ability to transmit multiple cues and typically contain fewer cues than richer mediums (Lengel & Daft, 1988). Conversely, face-to-face communication is considered the richest medium because it can transmit multiple cues and information at once (Lengel & Daft, 1988). Richness of the channel is dependent upon the ability to communicate information, the ability to handle multiple cues, feedback rate, and the amount of personal focus (Lengel & Daft, 1988). These factors may also be the reason why some channels would be more appealing to teleworkers of differing personality types such as email, instant messaging, and video communication.

Channel Expansion Theory

Channel expansion theory was conceptualized in an attempt to reconcile previous media richness research (Carlson & Zmud, 1999). While media richness theory has generally been supported when tested on traditional media such as face-to-face and phone communication, the findings have been inconsistent about new media such as e-mail (Lengel & Daft, 1988). Thus, the central premise of channel expansion theory is that an individual's experiences are important factors influencing a person's perception of channel richness. Channel expansion theory argues that each person develops a richness perception for communication channels, specifically influenced by four experiences: experience with the channel, experience with the messaging topic, experience with the organizational context, and experience with communication co-participants (Carlson & Zmud, 1999). Increases in these four types of experience should allow people to articulate and recognize indicators that signal rapid feedback, multiple cues, natural language, and personal focus. For example, co-workers who frequently communicate via email will become more aware of how to convey different levels of formality and communicate subtleties with more experience. Therefore, these types of experiences are positively related to a person's perception of a channel's richness (Carlson & Zmud, 1999).

Similar to how people develop experience with a channel, they also develop experience with communication partners, such as co-workers and supervisors. As people communicate with a specific communication partner, they begin to develop a knowledge base for that person, allowing them to communicate messages tailored to their partner making for a richer communication experience. This can be accomplished through using cues relevant to him or her, referring to shared experience, or using common language (Carlson & Zmud, 1999). This type of knowledge is acquired through on-going communication and the use of one or more knowledge-generating strategies to develop knowledge about others (Walther, 1996). Additionally, as people develop experience with a communication topic, they develop a knowledge base for the topic, allowing for richer communication experiences (Carlson & Zmud, 1999). When communication partners have similar topic experience, richer messages can be facilitated through leveraging shared understanding. As such, communication partners can interpret messages received about a topic more or less richly based on their topic knowledge (Carlson & Zmud, 1999). Finally, people develop a knowledge base centered upon the organizational context in which they are communicating. This allows for communication partners to encode messages with shared symbols and/or organizational cultural references for a richer communication experience, although this idea has found only partial support in empirical research (Carlson & Zmud, 1999).

While experience with communication partners, knowledge, and organizational context are essential to perceptions of media richness, these concepts alone do not fully explain channel perceptions. The social influence model of technology use refers to individual beliefs concerning the appropriate use of a channel as well as perceptions of a channel's richness and demonstrates that these perceptions are in part socially constructed and therefore subject to social influence (Carlson & Zmud, 1999). Research about the social influence model has engendered mixed results. For example, in Carlson and Zmud's two-wave study, support was only found in one group. Although rationalized as a research design error, this was not the first study to find a lack of support for the social influence model (Schmitz & Fulk, 1991). However, D'Urso and Rains' (2008) study did find that perceptions of media richness are socially constructed, which is in line with the social influence model.

9.1.2 Communication Satisfaction and Generational Differences: Our Data

Downs and Hazen's (1977) *Communication Satisfaction Scale* was implemented to index the degree of satisfaction with communication within one's organization. This scale contains five subdimensions, including: personal feedback, organizational identification, communication climate, horizontal communication, and relationship with supervisor. Overall means for the main scale and all subscales were densely concentrated above the midpoint, ranging from 3.97 (relationship with supervisor) to 3.61 (communication climate). Differences across generations are presented below in Table 9.1.

Table 9.1 Generational Differences in the Communication Satisfaction Scale and Subscales

Measure	*Baby Boomers* M(SD)	*Gen X* M(SD)	*Millennials* M(SD)	F (*2, 1147*)	*eta*^{2}
Communication Satisfaction	3.94(.76)	3.77(.86)	3.76(.81)	1.989	.00
Personal Feedback	3.82(1.00)	3.62(1.07)	3.64(.99)	1.358	.00
Organizational Integration	**4.12(.80)**[a]	3.89(.94)[a,b]	**3.81(.88)**[b]	5.575*	.01
Communication Climate	3.72(.99)	3.55(1.03)	3.61(1.00)	.963	.00
Horizontal Communication	3.86(.73)	3.81(.78)	3.83(.77)	.148	.00
Relationship with Supervisor	**4.19(.91)**[a]	3.96(1.03)[a,b]	**3.93(.96)**[b]	3.049*	.01

**$p < .001$, *$p < .05$.

[a,b]Significant differences between groups as determined by Tukey HSD post hocs. Scores which significantly differ from the other two scores are bolded.

The results of the between-groups comparisons below reveal minor differences between some generational groups on two of the subscales. Overall, the results show a high degree of communication satisfaction. Two subscales see Baby Boomers reporting significantly greater satisfaction in: *organizational integration*, which is concerned with the degree to which employees receive information that is pertinent to their immediate responsibilities, and greater satisfactions with their relationship with their supervisors. Baby Boomers differ only from Millennials on these two scales, as Gen X falls squarely in between each group and does not differ from either.

This is a particularly useful scale to measure communication satisfaction because it examines the different dimensions that work together to create communication satisfaction. It is not surprising that Baby Boomers are higher on organizational integration. This could be explained through another factor, such as length of time employed by the organization. Even if that did not explain this finding, overall level of work experience could. Since Baby Boomers have been a part of the workforce for a longer period of time than other generations in this study, they have likely learned how to integrate faster and evaluate their integration differently than they did earlier in their career. Conversely, this could help to explain why there is a stagnation among Gen X. Early in their careers, they probably experienced a "honeymoon" phase of employment when they were learning and happy to be employed and soaking everything in, trying to understand how they would evaluate workplace communication. Then, as their experience grew, their perceptions changed because they had baseline knowledge, which could cause their perceptions of communication satisfaction, in this instance, to drop or plateau. This is likely why our findings show Gen X to be lower on some dimensions.

9.2 Generational Perspectives on Job Satisfaction

Job satisfaction refers to how content an individual is with his or her job (West & Berman, 2009). Job satisfaction can also be defined as an affective relationship to one's job that is a function of situational factors, including nature of the work, human resources elements, and the organizational environment (Boswell, Shipp, Payne, & Culbertson, 2009). Previous research about job satisfaction and age remains inconclusive. Therefore, it is unclear if there is a link between job satisfaction and generational membership.

One of the biggest reasons why there is inconsistency in research about age and job satisfaction is because job satisfaction remains stable over time for many employees. However, older employees tend to either maintain their level of job satisfaction over time or experience a decrease in job satisfaction because the number of opportunities available to them decreases (Applebaum, Serena, & Shapiro, 2005). In a study of hospitality workers, generational differences between Millennials and Baby Boomers were found to have significant moderating effects on the relationship between emotional exhaustion and job satisfaction and turnover intention (Lu & Gursoy, 2013). In a comprehensive study conducted by the Society for Human Resource Management, (SHRM, 2014), it was reported that most Millennials are satisfied with their job and that compensation and job security are the two main contributors to their evaluation and feelings of job satisfaction. They also feel satisfied with their job when they are motivated by their work goals, which encourages leaders to help Millennial employees set goals related to their job. Members of Gen X, however, tend to report the lowest levels of job satisfaction because they feel like they are not rewarded enough for their hard work and dedication to their organizations (MetLife, 2019).

9.2.1 Previous Research and Theory

As previously mentioned earlier in this chapter, job satisfaction is a multidimensional construct that is influenced by several other factors, including, but not limited to: organizational culture, communication, leadership, personality, and teleworking. This section will briefly outline some of the previous research findings related to job satisfaction, with the note that this is not an exhaustive overview of this multidimensional construct that is a robust area of research.

An early model of job satisfaction research proposed the two-factor model (Herzberg, Maunser, & Snyderman, 1959), which later gave way to the global approach for understanding job satisfaction. The global approach studies separate job parts that are likely to promote or prevent an individual's level of job satisfaction (Sowmya & Panchanatham, 2011). This informed another model of job satisfaction by Hackman and Oldman (1975), which considered five variables to create "motivating potential," which is the degree to which an employee's motivation can be influenced. Over time, researchers turned to more cognitive conceptualizations of job satisfaction, which is where the field remains today, examining variables that include the employee's needs and how they perceive job satisfaction.

Through cognitive research, we have learned job satisfaction is dependent on many things beyond an employee's frame of mind and also influenced by organizational factors like culture, size, salary, and working conditions or environment. For example, there is a large body of work that examines the job satisfaction of teleworking employees, which has become a renewed interest due to the pandemic that occurred in 2020 and forced organizations to allow many employees to work remotely.

A study conducted in 2005 found a curvilinear relationship between teleworking and job satisfaction, arguing that at a certain point, teleworking leads to less employee satisfaction (Golden, Veiga, & Simsek, 2006). Specifically, Golden and Viega argued that in small amounts, teleworkers are more satisfied because they experience all the benefits of teleworking, while minimizing the disadvantages of teleworking, such as isolation and lack of interpersonal workplace relationships. However, when employees telework regularly, or exclusively, they may start to experience the disadvantages more heavily than the advantages, and therefore have less job satisfaction. Teleworking initially, and in smaller amounts, increases job satisfaction due to its many benefits. Conversely, as the level of teleworking increases and becomes more frequent or exclusive, job satisfaction decreases and at some point, plateaus. These results show that teleworking is complex, and job satisfaction can vary for a number of different reasons, such as the amount of time-spent teleworking (Golden et al., 2006).

In contrast to the study conducted by Golden et al. (2006), a meta-analysis suggested a positive relationship between teleworking and job satisfaction (Gajendran & Harrison, 2007) by demonstrating positive direct effects between teleworking and job satisfaction, mostly attributed to the many benefits teleworking provides employees. In fact, in contrast to previous findings, the meta-analysis did not find any negative effects of teleworking on workplace social ties. Indeed, the meta-analysis revealed that while there are disadvantages to teleworking, the advantages and benefits of teleworking outweigh the disadvantages when it comes to job satisfaction but did not account for the possibility communication channel satisfaction may play in the teleworking and job satisfaction relationship. Finally, Smith, Patmos, and Pitts (2018) found that personality composition does affect the relationship between teleworking and job satisfaction; for example, those who are higher in extraversion, openness, agreeableness, and conscientiousness are best suited for job satisfaction in teleworking environments.

Another major area of inquiry is the relationship between job satisfaction and leadership. Unsurprisingly, employees are more satisfied in organizations that are flexible and that emphasize communication and reward employees (McKinnon, Harrison, Chow, & Wu, 2003), which is greatly influenced by leadership style. Within job satisfaction research, two types of leadership have been studied: transactional leadership and transformational leadership. Transactional leaders act within existing organizational culture, wheres transformational leaders initiate change and can adapt the organizational culture to their own values (Belias & Koustelios, 2014). When leadership style is a match to the employee's values and vision, job satisfaction is positively influenced (Chang & Lee, 2007). Other research has supported the idea that transformational leadership can lead to greater job satisfaction (Bushra, Usman, & Naveed, 2011; Emery & Barker, 2007; Riaz, Akram, & Ijaz, 2011).

9.2.2 Job Satisfaction and Generational Differences: Our Data

Job Satisfaction was measured with a six-item, unidimensional scale from Pond and Geyer (1991). The scale demonstrated excellent reliability, Cronbach's alpha = .966. Greater scores on this measure indicate greater satisfaction.

Although the mean score on Job Satisfaction for Baby Boomers (M = 3.96, SD = 1.09) was slightly higher than both Gen X (M = 3.71, SD = 1.81) and Millennials (M = 3.75, SD = 1.02), that difference failed to reach statistical significance (F(2, 1147) = 1.957, p = .142, partial eta squared = .00). This pattern of findings, where Baby Boomers' raw scores on job satisfaction are higher, but not significantly so, is similar to the findings reported for Communication Satisfaction above. The results presented in this chapter note a couple of areas where Baby Boomers are slightly more satisfied with communication in their organizations, but no significant differences emerged in overall communication satisfaction of job satisfaction across the three generations explored here.

For similar reasons, it is not surprising that there are not statistically significant generational differences within job satisfaction. Knowing that job satisfaction is a multidimensional construct, this information can be used to help inform future research to better determine, where, if at all, generational differences might influence job satisfaction. Our findings also demonstrate that there is a lot of room to further improve job satisfaction and make it even stronger, through initiatives that enhance or maintain workplace communication.

9.3 Best Practices for Creating a Satisfying Workplace

This chapter has demonstrated the interrelatedness of communication satisfaction and job satisfaction with several other workplace variables including organizational culture, support, leadership, and individual differences. Therefore, the best practices for creating a satisfying workplace among multigenerational employees reflects the relationship that these constructs have with others.

Best Practice 1: Consider Culture

As previous research indicates, organizational culture is really central to understanding satisfaction. If issues exist with constant miscommunications or high levels of communication dissatisfaction, there is most likely a cultural block somewhere within the organization. Similarly, if the retention rate of employees is low, morale is low, and you know that a lot of employees are dissatisfied for a prolonged period of time, you need to look to the organizational culture to address the problem and make changes that promote a culture of satisfaction. This is best measured by looking at employee satisfaction surveys and analyzing exit interview data to understand why people are leaving and if there is any opportunity to make changes that can increase retention and overall satisfaction. The fact is, satisfied employees stay with organizations, so turnover rate is a key metric for understanding employee satisfaction.

Best Practice 2: Communication is Central

The information within this chapter also demonstrated the central role of communication. Although culture is vital to understanding satisfaction, culture is only shared through various forms of communication. When communication is poor throughout an organization, job satisfaction will likely be lower. Conversely, when communication is satisfying and employees feel heard, informed, and able to easily share information, they are likely to report greater levels of both communication and job satisfaction. Refer back to the information in Chapter 7 about supportive communication to improve employee communication satisfaction, which does not seem to have statistically significant generational differences.

Best Practice 3: Technology is a Tool

Technology has overtaken communication and the workplace. Previous research does indicate that there are generational differences related to technology in the workplace, and this is an important consideration with regard to communication and job satisfaction. Generational differences may be present among teleworkers when it comes to job satisfaction.

Consistently using inappropriate channels to share information can lead to lower levels of communication satisfaction, which over time can have a negative and cumulative effect on job satisfaction. Therefore, organizations need to be mindful about the use of technology and not only create policies that promote healthy and satisfactory use but also encourage leaders to model this behavior accordingly to better demonstrate and build satisfaction levels among a multigenerational workforce.

9.4 Conclusion

Communication satisfaction and job satisfaction are highly correlated for every workforce, not just a multigenerational one. Therefore, it is important to understand the different variables that can positively and negatively impact both, which can be influenced by generational membership. The findings from our study are descriptive and demonstrate that when satisfaction is present, there are negligible if any, generational differences among employees. Therefore, this should be encouraging news for organizations and leaders to promote effective communication in the workplace for long-term job satisfaction.

References

Akkirman, A. D., & Harris, D. L. (2005). Organizational communication satisfaction in the virtual workplace. *Journal of Management Development*, *24*, 397–409.

Applebaum, S., Serena, M., & Shapiro, B. (2005). Generation "X" and the Boomers: An analysis of realities and myths. *Management Research News*, *28*, 1–32.

Belias, D., & Koustelios, A. (2014). Organizational culture and job satisfaction: A review. *International Review of Management and Marketing*, *4*, 132–149.

Boswell, W., Shipp, A., Payne, S., & Culbertson, S. (2009) Changes in newcomer job satisfaction over time: Examining the pattern of honeymoon and hangovers. *Journal of Applied Psychology*, *94*, 844–858.

Bushra, F., Usman, A., & Naveed, A. (2011). Effect of transformational leadership on employee's job satisfaction and organizational commitment in banking sector of Lahore (Pakistan). *International Journal of Business and Social Science*, *2*, 261–267.

Carlson, J., & Zmud, R. (1999). Channel expansion theory and the experiential nature of media richness perceptions. *Academy of Management Journal*, *42*, 153–170.

Chang, S., & Lee, M. (2007). A study on relationship among leadership, organizational culture, the operation of learning organization and employee's job satisfaction. *The Learning Organization*, *14*, 155–185.

Crino, M. D., & White, M. C. (1981). Satisfaction in communication: An examination of the Downs-Hazen measure. *Psychological Reports*, *49*, 831–838.

Downs, C. W., & Hazen, M. D. (1977). A factor analytic study of communication satisfaction. *Journal of Business Communication*, *14*, 63–73.

D'Urso, S., & Rains, S. (2008). Examining the scope of channel expansion: A test of channel expansion theory with new and traditional communication media. *Management Communication Quarterly*, *21*, 486–507.

Easton, S., & Bommelje, R. (2011). Interpersonal communication consequences of email non-response. *Florida Communication Journal*, *39*, 45–63.

Emery, C., & Barker, K. (2007). The effect of transactional and transformational leadership styles on the organizational commitment and job satisfaction of customer contact personnel. *Journal of Organizational Culture, Communications, and Conflict*, *11*, 77–90.

Gajendran, R., & Harrison, D. (2007). The good, the bad, and the unknown about telecommuting: Meta-analysis of psychological mediators and individual consequences. *Journal of Applied Psychology*, *92*, 1524–1541.

Golden, T., Veiga, J., & Simsek, Z. (2006). Telecommuting's differential impact on work-family conflict: Is there no place like home? *Journal of Applied Psychology*, *91*, 1340–1350.

Green, M., Hilken, J., Friedman, H., Grossman, K., Gasiewski, J., & Adler, R. (2005). Communication via instant messenger: Short and long-term effects. *Journal of Applied Social Psychology*, *35*, 445–462.

Hackman, J., & Oldman, G. (1975). Development of the Job Diagnostic Survey. *Journal of Applied Psychology*, *60*, 159–170.

Henderson, S., & Gilding, M. (2004). "I've never clicked this much with anyone in my life": Trust and hyper-personal communication in online friendships. *New Media & Society*, *6*, 487–506.

Herzberg, F., Maunser, B., & Snyderman, B. (1959). *The motivation to work*. John Wiley & Sons, Inc.

Hollingshead, A., McGrath, J., & O'Connor, K. (1993). Group task performance and communication technology: A longitudinal study of computer-mediated versus face-to-face work groups. *Small Group Research*, *24*, 307–333.

Jacobsen, D. (1999). Impression formation in cyberspace: Online expectations and offline experiences in text-based virtual communities. *Journal of Computer-Mediated Communication*, *5*. doi: 10.1111=j.1083–6101.1999.tb00333.x

Kock, N. (2004). The psychobiological model: Towards a new theory of computer-mediated communication based on Darwinian evolution. *Organizational Science*, *15*, 327–349.

Kurland, N., & Bailey, D. E. (1999). Telework: The advantages and challenges of working here, there, anywhere, and anytime. *Organizational Dynamics*, *28*, 53–68.

Lengel, R., & Daft, R. (1988). The selection of communication media as an executive skill. *The Academy of Management Executive*, *2*, 225–232.

Lu, A., & Gursoy, D. (2013). Impact of job burnout on satisfaction and turnover intention: Do generational differences matter? *Hospitality & Tourism Research*, *40*, 210–235.

McKinnon, L., Harrison, L., Chow, W., & Wu, A. (2003). Organizational culture: Association with commitment, job satisfaction, propensity to remain and information sharing in Taiwan. *International Journal of Business Studies*, *11*, 25–44.

MetLife (2019). Thriving in the new work-life world. MetLife's 17th annual U.S. employee benefit trends study. https://www.metlife.com/content/dam/metlifecom/us/ebts/pdf/MetLife-Employee-Benefit-Trends-Study-2019.pdf

Park., H. S., & Raile, A. N. W. (2010). Perspective taking and communication satisfaction in coworker dyads. *Journal of Business Psychology*, *25*, 569–581.

Pincus, J. D. (1986). Communication satisfaction, job satisfaction, and job performance. *Human Communication Research*, *12*, 395–419.

Pond, S., & Geyer, P. (1991). Differences in the relation between job satisfaction and perceived work alternatives among older and younger blue-collar workers. *Journal of Vocational Behavior*, *39*, 251–262.

Riaz, T., Akram, M., & Ijaz, H. (2011). Impact of transformational leadership style on affective employees commitment: An empirical study of banking sector in Islamabad (Pakistan). *The Journal of Commerce*, *3*, 43–51.

Schmitz, J., & Fulk, J. (1991). Organizational colleagues, media richness, and electronic mail: A test of the social influence model of technology use. *Communication Research*, *18*, 487–523.

Sheer, V. (2011). Teenagers' use of MSN features, discussion topics, and online friendship development: The impact of media richness and communication control. *Communication Quarterly*, *59*, 82–103.

Sheer, V., & Chen, L. (2004). Improving media richness theory. *Management Communication Quarterly*, *18*, 76–93.

SHRM (2014). Millennial employees' job satisfaction and engagement. *SHRM Research Spotlight*. Retrieved from: https://www.shrm.org/hr-today/trends-and-forecasting/research-and-surveys/Documents/Millennials-Job-Satisfaction-Engagement-Flyer.pdf

Sias, P. M., Pedersen, H., Gallagher, E. B., & Kopaneva, I. (2012). Workplace friendship in the electronically connected organization. *Human Communication Research*, *38*, 253–279.

Smith, S. A., Patmos, A. K., & Pitts, M. J. (2018). Communication and teleworking: A study of communication channel satisfaction, personality, and job satisfaction for teleworking employees. *International Journal of Business Communication*, *55*, 44–68. doi: 10.1177/2329488415589101

Sowmya, K., & Panchanatham, N. (2011). Factors influencing job satisfaction of banking sector employees in Chennai, India. *Journal of Law and Conflict Resolution*, *3*, 76–79.

Tsai, M., & Chuang, S. (2009). An integrated process model of communication satisfaction and organizational outcomes. *Social Behavior and Personality*, *37*, 825–834.

Walther, J. (1996). Computer-mediated communication: Impersonal, interpersonal and hyperpersonal interaction. *Communication Research*, *23*, 3–43.
Watson-Manheim, M. B., & Belanger, F. (2007). Media repertoires: Dealing with the multiplicity of media choices. *MIS Quarterly*, *31*, 267–293.
West, J., & Berman, E. (2009). Job satisfaction of public managers in special districts. *Review of Public Personnel Administration*, *29*, 327–353.

10 Remote and Virtual Work

This chapter describes the changing nature of the workplace specifically focusing on remote and virtual work. The evolving nature of work, especially surrounding digital expectations, will be discussed.

10.1 The Evolving Nature of Work

At the time of this writing, the very nature of work has changed dramatically on an international scale. On December 31st, 2019, China confirmed an unknown cause for a handful of pneumonia cases. By the first week of January 2020, this unknown disease had been labeled as a new strain of coronavirus effectively named COVID-19. Travel bans, economic declines, event cancellations, lockdowns, and stay-at-home orders ensued as the virus spread worldwide and the death toll rose.

In essence, this new coronavirus strand impacted everything, especially work. From an economic standpoint, COVID-19 created unprecedented "modern-era" job loss. In April 2020, unemployment in the United States reached upward of 14%, or around 23 million unemployed workers. Some workers were furloughed with subsequent layoffs days, weeks, or months after. Not all jobs lost could have been performed in a remote capacity; however, many jobs were saved because of a virtual transition.

Though the COVID-19 pandemic of 2020 ushered forth a new era wherein remote and virtual work became more prominent, remote and virtual work are not new phenomena. Even as early as 1989, Peter Drucker supposedly determined that working from the office was obsolete (Streitfield, 2020). Mokhtarian (1991) defined telecommuting, broadly, as the use of "telecommunications technology to partially or completely replace the commute to and from work" (p. 1). This distinction, developed by Mokhtarian 30 years ago, has only somewhat changed. Remote work today still encompasses a work-from-somewhere-else mantra, but the National Institute of Health (NIH) does distinguish remote work from telework among their own workers. According to National Institutes of Health (2020), remote work constitutes a permanent designation in which the employee works permanently from an alternative worksite, while telework is regular or ad hoc but does require the employee to report to the office at times. These distinctions are helpful for considering next steps in terms of defining the structure of work for employees. Semantics, especially now, are crucial as organizations determine how to proceed for telecommuting or remote workers. No matter how those distinctions are communicated organizationally, the overall sentiment remains the same: work not consistently conducted in the physical office setting.

There are, generally, different types of remote work and subsequently different remote workers. Nickson and Siddons (2012) identify a home worker as anyone who is based at home and who uses their home as their main office space for at least two days a week. Home office working, though, is different from coworking, where individuals still work at a physical office location but work alongside other unaffiliated professionals (Spinuzzi, 2012). Coworking, post pandemic, has become less popular, and the term "remote work" generally applies to a virtual context. With that said, a more helpful all-encompassing term was coined by Grant, Wallace, and Spurgeon (2013) who refer to remote e-working as work completed anywhere and at any time. This reality, that work can legitimately be accomplished anywhere regardless of location, presents incredible opportunities and challenges for organizations today.

Workers in remote contexts experience a different work environment from the traditional face-to-face worker. For one, they have limited face-to-face interaction with their peers or supervisors (Charalampous, Grant, Tramontano, & Michailidis, 2019) which can have negative effects on support and employee engagement. In addition, the actual environment where remote workers conduct their business can lead to home conflict and distraction (Eddleston and Mulki, 2017). These issues create potential difficulty for remote workers, but these challenges can be overcome.

The modern workplace, highlighted by shifts in remote and virtual work, has changed in other ways as well. The delivery of work, or how we conduct our work and in what context, is not the only issue to consider. The actual work done has also become a topic of discussion. The virtual work revolution has provided an opportunity for work "gigification" (Veen, Kaine, Goods, & Barratt, 2020). This means, basically, that new services and novel solutions to niche problems have been answered through many different platforms or opportunities. Generally, "gig" work is driven by customer demand

and the individual worker, typically, has what is needed to perform the work. Usually "gig" work is paid at a piece rate (think freelance), and there is usually an intermediary platform that connects client and worker. The ability to not just work for a full-time employer but also conduct business virtual for "gig" platforms has placed a premium on remote and virtual work contexts.

This evolution of work makes leadership all the more important. Flood (2019) believes, and rightly so, that leading a remote or virtual workforce requires formal policies, additional resources, and innovative operations for individuals to share ideas and achieve objectives. Thus, performance management, relationship building, communication, and training all become even more important in a virtual or remote context (Hickman & Pendall, 2018). As remote work continues to become more popular in a "post-pandemic" world, leaders need to be adept not only at managing generational differences but also doing so in potentially remote and virtual environments.

10.2 Work Post-Pandemic

Since the pandemic, an unprecedented number of employers have begun working from home. In July 2020, job losses that occurred as a result of the COVID-19 pandemic were up to three times as large for non-remote workers (Angelucci, Angrisani, Bennett, Kapteyn, & Schaner, 2020). Not surprisingly, income losses and health challenges were greater during this time for non-remote workers. In addition, similar to the changes wrought by the gig-economy, the actual work conducted in a post-pandemic world varies. Bartik, Cullen, Glaeser, Luca, and Stanton (2020) report that remote work levels are higher post-pandemic but that there are significant variations across industries and that remote work is more common in industries with workers who are better educated and have better pay.

The final verdict on the actual effect and impact of remote work continues to be a matter of debate. Some, like Ozimek (2020), suggest that the remote work experiment has done better than expected. However, as will be discussed later, other dimensions related to productivity have yet to come to true fruition (Chebly, Schiano, & Mehra, 2020). Almost certainly, organizations will continue to figure out how to increase effectiveness and productivity in the remote workplace. However, as organizations strive to create environments that are flexible and adaptable for a new generation of employees, the underlying necessity to make operations, policies, and procedures remote friendly continues to confront remote leaders and workers.

10.3 Remote Work and Virtual Effectiveness

In a time that now seems like ages ago, Marissa Mayer, then CEO of Yahoo, banned remote work (Calvo, 2013). Mayer's decision was nothing short of controversial; today it would seem outrageous. Even before Millennials infiltrated the workplace, the concept of remote work and the ability to finish tasks virtually was debated but did find friendly audiences in several different industries. As Millennials became more ingratiated in the workplace, the demands for remote and virtual opportunities increased (Calvo, 2013). This reality has become even more prominent in our post-pandemic world. However, it is important to consider the benefits as well as the challenges of working in a remote or virtual context.

Benefits of Remote Work

It is important to note, first, that remote work can be successful because of great technological connectivity (Messenger & Gschwind, 2016). Without new technological advances, remote work would cease to exist, much less serve as a tangible alternative to face-to-face office work. New remote work tools and platforms are discussed later in the chapter, but before having a philosophical discussion about the merits of virtual work, it is important to note that as the technology goes, so goes the ability to do what employees need to do outside of the office.

The benefits of remote work are multifaceted and help more than just the individual employee. For one, remote work can decrease costs associated with purchasing a physical office or work location as well as maintaining that physical space (Felstead & Henseke, 2017). This reality helps affirm at least one potential organizational benefit. The necessary physical resources diminish as employees spend more time working from a non-office location. Kylili et al. (2020) go further and state that the lack of commute and necessity to be in a physical location also positively impacts the environment. For instance, decreased commuting leads to fewer emissions and a reduced carbon footprint. Additional environmental benefits include reduced noise pollution, reduced need for roads and infrastructure, and reduced road congestion (Kylili et al., 2020).

The concept of work-from-home, or today's even more appropriate mantra of work-from-anywhere (Choudhury, Foroughi, & Larson, 2020), offer employees great flexibility. This flexibility, however, must be managed appropriately. Staples, Hulland, and Higgins (1999) noted increased perceptions of employee productivity when their self-efficacy and perceptions of remote work effectiveness are high. This is important to note. Employees may not naturally find themselves astute at remote work operations, especially if the concept or experience is new, but organizations can employ training mechanisms to help employees understand expectations. Interestingly, Baker, Avery, and Crawford (2007) found that organizational and job-related factors are more likely to affect the satisfaction of work from home or work from anywhere employees. Meaning, it is less about household characteristics and work styles and still, like traditional face-to-face work contexts, more about the job and the culture itself.

Continually, increasing work life balance is a strong argument for remote work offerings. A better or more balanced work life environment can certainly be achieved if people work from home (Kerslake, 2002). The time saved by forgoing a commute and potentially saving on childcare are extremely beneficial. Yet, working virtually does present opportunities for overwork.

Potential Challenges of Remote Work

The concept of overwork, or the inability of employees to separate work from life on an even more extreme basis, is one of the most pressing questions, and challenges, of remote work. While work life balance in some ways is enhanced through remote and virtual work, the ability for employees to work too much is a real concern (Grant et al., 2013). Working from home does create challenges as employees try to separate their work from their regular life without clear differentiating boundaries.

In addition, the actual office environment, and the lack of social relationships at work can be problematic. Workers have said they miss office interactions and have a sense and feeling of isolation (Grant et al., 2013; Mann & Holdsworth, 2003). This isolation, which can be overcome, somewhat, with real-time, synchronous video or phone calls, is still a real manifestation of a potentially lonely situation.

Virtual team collaboration, creativity, and overall productivity may also suffer. Eisenberg and Krishnan (2018) identify five virtual team challenges, including trust and relationships, communication and knowledge sharing, perceptions and decision making, leadership, and diversity. These five challenges do not occur solely in a virtual context, but it is still worth noting that they can potentially become more pressing when workers are only interacting and communicating through remote platforms.

Finally, human resources professionals note issues related to overall employee clarity. For one, a general lack of policies and procedures are in place (Flood, 2019) across the spectrum as organizations try to identify the best ways to engage virtual workers while holding them accountable. The problems with work life balance and the ability to work at all times throughout the day prove especially problematic for hourly workers, who, in many instances, cannot go over a certain time threshold without incurring required overtime pay. These challenges are not insurmountable but certainly should cause organizations to consider best practices for engaging and managing remote employees.

10.4 Remote Work Tools and Platforms

The platforms used to perform remote work continue to provide new avenues for increased productivity; however, that was not always the case. The 1970s brought to the forefront an explosion of personal computers and, eventually, work-from-home monitoring software. The tool itself, the personal computer, was somewhat nullified without the ability for organizations to hold employees accountable through systemic network software. The trend continued with the advent of the Internet and the incredible capabilities the world wide web afforded. In the 1990s, the invention of WIFI provided greater geographic flexibility as workers could forsake their hard-wired Internet connections. After these foundational developments, specifically network and tools, the creation of monitoring software and actual remote work platforms started to increase. Eventually, with the creation of the smartphone, all roads converged into uber remote convenience.

From a policy standpoint, several governmental actions in the United States popularized or helped normalize remote work. For one, in 1995, Congress approved permanent funding for "flexiplace" work-related equipment for federal employees. Then in 2000, the DOT Appropriations Act was enacted. This law required all executive agencies to develop telecommuting policies. Five years later, in 2005, President Barack Obama signed the Telework Enhancement Act requiring federal agencies to create policies for employees to work remotely. As these policies were enacted, the actual number of remote workers and remote platform users continued to grow.

In 1987, 1.5 million documented telecommuters worked across the USA. Before the pandemic of 2020, 7 million people reported working remotely in the United States, roughly 3.5% of the population. In 2016, the team collaboration tool, Slack, had 4 million active daily users and 9 million by 2019. By 2018, 70% of the world's population reported working

Table 10.1 Remote Work Tools, Use, and General Purpose

Remote Work Tool	*General Use*	*Purpose*
Slack	Team communication application	Get instant feedback and connect with colleagues
Zoom	Video conferencing application	"Zoom calls" allow employees to meet synchronously over video
Trello	Project management software	Uses visual cues and aesthetic design to increase remote group productivity
Google Drive	File management application	Digital file management tool that allows for synchronous collaboration and file storage
Zapier	Workflow automation	Zapier allows applications to connect somewhat seamlessly to allow greater functionality and moves data between different applications

remotely at least once a week. By 2019, the team video collaboration tool, Zoom, reported more than 50,000 customers that had more than 10 employees. In 2017, Zoom reported more than 700,000 users. These platforms continue to grow in popularity and overall efficiency.

The market today is flooded with remote work platform options. For simplicity, we developed a table to summarize different remote work platforms. The list below is not exhaustive and mentions tools primarily popular in 2020 but does provide a glimpse into the varying types of workflow platforms and their focus (Table 10.1).

The list above does not account for duplicate products, like Skype or previous Google Hangouts for video conferencing, all-in-one platforms like Microsoft Teams, or even additional tools like Serene, Toggl, Spark, or remote desktop that allow for even more creativity and efficiency. These tools all can be helpful, but employers would be wise to think strategically about communication in their remote work environment and approach these tools with intentionality and purpose. Just because a tool exists does not mean it is best for every business or organization.

10.5 Generational Differences and Remote Work

As remote work becomes more popular, and remote work tools become more widespread, a look at how employees feel about remote and virtual work can provide organizational clarity. While not the only framework to consider employee perceptions of remote work, generational differences can help supervisors and executives determine how their employees differ in their remote work perceptions. With that said, the overall differences may not be as distinct as one would assume, especially in a world overrun with remote workers and remote work opportunities.

Before considering legitimate generational differences in this regard, it is worth noting again what different generations value. Millennials, for instance, are tech-savvy and appreciate collaborative work, Gen Xers tend to rank lower on executive presence, and Baby Boomers are notably loyal to their organizations but may not adapt quickly. In addition, like Millennials, Gen Xers appreciate flexibility and want to be independent workers. Zimmerman (2016) reports that Millennials appreciate working from home, but an even greater number of Boomer and Gen X employees prefer a work from home option. Millennials are generally used to the always-available workplace, yet they continually rank work-life balance and work flexibility as highly-rated work rewards.

Kennelly (2015) correctly identifies that new employee characteristics are shifting in terms of when people work, how people share jobs, and how individuals are evaluated on their performance. However, even as these shifts occur, employees generally desire some sort of flexibility in terms of how their job is performed (Rousseau, 2015). This trend is probably not going away.

Recognizing that all generations prefer flexibility, in some capacity, all workers may not be prime candidates for work-from-home or work-from-anywhere options. As organizations continue to explore remote work options, individual employee desires; the actual job description and primary responsibilities; and remote work capacity should all be considered.

10.6 Best Practices

Best Practice 1: Consider the Purpose of Remote Work

While it may be easy – especially in a work world initially ravaged by a pandemic – to assume jobs work best in a remote context, employers and supervisors should consider all facets of the job, the organizational culture, and individual employees before demanding or suggesting employees work from home. Organizations would do well to intentionally explore remote options without assuming individuals should or should not work from home because of personal preferences. This chapter does position remote work as a tangible and realistic option for a host of employees for a number of reasons. As a suggested best practice, though, organizations are encouraged to tread wisely into remote positions.

Best Practice 2: Remote Work Platforms Are NOT Created Equally

While the lists presented above are not exhaustive, they do provide a snippet of the software or platform options available to organizations. However, organizations should navigate platform choices wisely. For one, remember that all software platforms and remote network systems are not created equal. Your organization may not need, for example, all of the features Slack has to offer. Consider your overall internal communication strategy as it relates to remote work platforms. Instead of using all tools available and switching when a new tool arrives on the market, work in a constant state of assessment and analysis, asking how employees use their current software to communicate effectively and what other features are needed in order to conduct appropriate business.

Best Practice 3: Designate Remote Responsibilities as well as Policies and Procedures

The rush to remote work in 2020 highlighted a few issues in the general workplace. For example, when many organizations transitioned to remote work, they were not prepared from a human resources perspective. Specifically, a lack of policies regarding work conducted remotely led to a lack of accountability for both employees and managers in some organizations. Any organization desiring a remote transition should first assess its overall operations and designate what should be completed remotely. After this, designating remote responsibilities for individual employees is crucial. Finally, policies and general procedures for remote work should be clearly communicated to the organization as a whole.

10.7 Conclusion

This chapter presented the current state of remote and virtual work. While remote work may seem relatively new because of the steep increase in remote workers due to global events, the reality is that remote work has existed for decades. This reality should encourage organizations who can move remotely to do so. From a generational perspective, it is interesting to note that employees, generally, across the spectrum appreciate the opportunity to do their work in a flexible environment, remote or otherwise.

This desire transcends generational differences and has become a general desire of employees who believe working from home gives greater work-life balance and offers more flexibility. With that said, organizations should approach remote work wisely. Do not forsake what makes your organizational culture unique in an effort to satisfy the new remote work urge. Instead, take time to think deeply about a potential remote work transition and, where possible, transition slowly and with purpose.

References

Angelucci, M., Angrisani, M., Bennett, D. M., Kapteyn, A., & Schaner, S. G. (2020). *Remote work and the heterogeneous impact of COVID-19 on employment and health* (No. w27749). National Bureau of Economic Research.

Baker, E., Avery, G. C., & Crawford, J. (2007). Satisfaction and perceived productivity when professionals work from home. *Research and Practice in Human Resource Management, 15*, 37–62.

Bartik, A. W., Cullen, Z. B., Glaeser, E. L., Luca, M., & Stanton, C. T. (2020). *What jobs are being done at home during the COVID-19 crisis? Evidence from firm-level surveys* (No. w27422). National Bureau of Economic Research.

Calvo, A. J. (2013). Where's the remote? Facetime, remote work, and implications for performance management. *Cornell HR Review*. http://digitalcommons.ilr.cornell.edu/chrr/45/

Charalampous, M., Grant, C. A., Tramontano, C., & Michailidis, E. (2019). Systematically reviewing remote e-workers' well-being at work: A multidimensional approach. *European Journal of Work and Organizational Psychology, 28*, 51–73.

Chebly, J., Schiano, A., & Mehra, D. (2020). The value of work: Rethinking labor productivity in times of COVID-19 and automation. *American Journal of Economics and Sociology, 79*, 1345–1365. https://doi.org/10.1111/ajes.12357

Choudhury, P., Foroughi, C., & Larson, B. Z. (2020). Work-from-anywhere: The productivity effects of geographic flexibility. In *Academy of Management Proceedings* (Vol. 2020, No. 1, p. 21199). Academy of Management.

Eddleston, K. A., & Mulki, J. (2017). Toward understanding remote workers' management of work–family boundaries: The complexity of workplace embeddedness. *Group & Organization Management, 42*, 346–387.

Eisenberg, J., & Krishnan, A. (2018). Addressing virtual work challenges: learning from the field. *Organization Management Journal, 15*, 78–94.

Felstead, A., & Henseke, G. (2017). Assessing the growth of remote working and its consequences for effort, well-being and work-life balance. *New Technology, Work and Employment, 32*(3), 195–212.

Flood, F. (2019). Leadership in the remote, freelance, and virtual workforce era. In A. Farazmand (Ed.), *Global encyclopedia of public administration, public policy, and governance* (pp. 1–7). Springer. https://doi.org/10.1007/978-3-319-31816-5_3825-1.

Grant, C. A., Wallace, L. M., & Spurgeon, P. C. (2013). An exploration of the psychological factors affecting remote e-worker's job effectiveness, well-being and work-life balance. *Employee Relations*, *35*, 527–546.

Hickman, A., & Pendall, R. (2018). The end of the traditional manager. Gallup. https://www.gallup.com/workplace/236108/end-traditional-manager.aspx

Kennelly, J. (2015). Embracing the virtual workforce. *Human Resource Management*, 22–23.

Kerslake, P. (2002). The work/life balance pay-back. *New Zealand Management*, *49*, 28–31

Kylili, A., Afxentiou, N., Georgiou, L., Panteli, C., Morsink-Georgalli, P. Z., Panayidou, A., Papouis, C., & Fokaides, P. A. (2020). The role of remote working in smart cities: Lessons learnt from COVID-19 pandemic. *Energy Sources, Part A: Recovery, Utilization, and Environmental Effects*, *12*, 1–16.

Mann, S., & Holdsworth, L. (2003). The psychological impact of teleworking: Stress, emotions and health. *New Technology, Work and Employment*, *18*, 196–211.

Messenger, J., & Gschwind, L. (2016). Three generations of telework: New ICT and the (R)evolution from home office to virtual office. *New Technology, Work and Employment*, *31*, 195–208.

Mokhtarian, P. (1991). Defining telecommuting. *UC Davis: Institute of Transportation Studies*. https://escholarship.org/uc/item/35c4q71r

National Institutes of Health. (2020, November). Remote work. https://hr.nih.gov/working-nih/work-schedules/remote-work

Nickson, D., & Siddons, S. (2012). *Remote working*. Routledge.

Ozimek, A. (2020). The future of remote work. *Available at SSRN 3638597*.

Rousseau, D. (2015). *I-deals: Idiosyncratic deals employees bargain for themselves: Idiosyncratic deals employees bargain for themselves*. Routledge.

Spinuzzi, C. 2012. Working alone together: Coworking as emergent collaborative activity. *Journal of Business and Technical Communication*, *26*, 399–441. doi:10.1177/1050651912444070.

Staples, D. S., Hulland, J. S., & Higgins, C. A. (1999). A self-efficacy theory explanation for the management of remote workers in virtual organizations. *Organization Science*, *10*, 758–776.

Streitfield, D. (2020, June 29). The long, unhappy history of working from home. *New York Times*. https://www.nytimes.com/2020/06/29/technology/working-from-home-failure.html

Veen, A., Kaine, S., Goods, C., & Barratt, T. (2020). 2 The 'gigification' of work in the 21st century. In *Contemporary work and the future of employment in developed countries* (pp. 15–32). Routledge-Cavendish.

Zimmerman, K. (2016, October 14). Do millennials prefer working from home more than Baby Boomers and Gen X? Forbes. https://www.forbes.com/sites/kaytiezimmerman/2016/10/13/do-millennials-prefer-working-from-home-more-than-baby-boomers-and-gen-x/?sh=475f7eb42070

11 Innovation and Future Challenges

This chapter explores the future of work, including future challenges, expectations, and evolutions. The authors provide information about future generations (i.e. what can we learn from Generation Alpha) and what communication will look like in the workplace of the future.

11.1 The Future of Work

The future of work looks drastically different in a post-pandemic world. However, some innovations we have already implemented, like work-from-anywhere and remote or virtual work, will continue to become more commonplace. Where we conduct our work has been a consistent discussion for years – if not decades – but other realities will distinguish work in a post-2020 world from work environments that came before. This section will preview several new work initiatives that may help define the future of work. Generally, we may see a trend toward what Peters (2017) calls technological unemployment because of so many new technologies replacing traditional vocations.

Automation

As a work phenomenon, automation includes artificial intelligence, autonomous systems, and robotics. Workplace automation replaces or enhances workflow and varying processes through technology. As one could assume, automation offers several benefits, specifically cutting costs and increasing productivity. As a framework or overarching change agent for the modern workplace, automation is in the midst of a renaissance or a revolution. Interestingly, AI and automation experts call our current climate the fourth Industrial Revolution (Sako, 2020). This new revolution brings together artificial intelligence, automation, robotics, genetic engineering, and other technologies. Automation receives mixed reviews from economists and employees as some declare it replaces employment, considering that is what it is designed to do (Autor, 2015). However, Bessen (2019) points out that automation can actually increase opportunities for industry employment. Generally, automation causes employees to, at the very least, learn new skills and new occupations. While opinions vary widely regarding the impact of automation on jobs (Winick, 2018), some clarity waits on the horizon as more companies and industries move to automated processes. In some ways, the COVID-19 pandemic of 2020 forced companies to make decisions about their digital or automated operations. No matter what comes next, automation – and one of the core subsets of automation, artificial intelligence – are certainly a significant category surrounding the future of work and work innovation. From a communication standpoint, the influx of automation brings about several interesting questions related to human capabilities; the need for emotional intelligence and workplace relationships; and the general, overall impact of automation on work and organizational culture.

Artificial Intelligence

The actual definition of artificial intelligence (AI), especially as it relates to a manifestation in the workplace, has been debated, and, unfortunately, no widely accepted definition of artificial intelligence exists. Wang (2019) reiterates the confusion involved with trying to determine one specific AI definition; however, simply, AI includes a "system's ability to correctly interpret external data, to learn from such data, and to use those learnings to achieve specific goals and tasks" (Kaplan & Haenlein, 2019, p. 15). In a work context, AI is predicted to have even more of a holistic globally transformative impact on the economy (Howard, 2019).

AI can be extremely beneficial in the workplace. For one, AI can actually help determine which jobs can be completed by a machine (Sako, 2020). This can significantly increase efficiency and productivity and provides some clarity about what tasks require a human worker for completion. These efficiencies come with some difficulties, and it is incorrect to assume artificial intelligence will just solve all workforce problems. There are some challenges to consider. As AI becomes more popular, additional regulation may be needed (Haenlein & Kaplan, 2019). Humans may need to develop ethics and functional norms surrounding AI (Sako, 2020). Regardless, artificial intelligence will continue to be a significant factor in the future of work.

Inclusive Workplaces

Inclusive work – and diversity and inclusivity as corporate initiatives – have become more popular as organizations wrestle with their hiring procedures and operational biases. Organizations have sought to become more inclusive by creating an atmosphere of psychological safety (Carmeli, Brueller, & Dutton, 2009); establishing transparent hiring practices; and providing equal access to advancement, decision-making, professional development, and other resources (Shore, Cleveland, & Sanchez, 2018). Inclusive workplaces will not occur naturally and instead must be achieved with careful consideration and planning.

Inclusive best practices are numerous. Offerman and Basford (2014) believe that inclusive leaders should work to develop a talent pipeline and that they should confront subtle discrimination. In addition, diversity should be leveraged to increase business performance, and accountability systems and mechanisms should be established (Offerman & Basford, 2014). Employees and other senior leaders should be trained to carry out inclusive efforts. Similarly, Sabharwal (2014) believes inclusive behaviors must be a top down initiative involving fair and equitable treatment of employees so they can influence organizational decisions. As a feature of the modern workplace and an integral part of the future of work, inclusive workplaces and inclusive organizational culture will continue to be an area of consistent study for communication scholars (Rezai, Kolne, Bui, & Lindsay, 2020).

Cloud-Based Collaboration

Cloud-based collaboration and cloud use in general will continue to be a feature of work moving forward. Gallaugher (2014) refers to cloud computing as hardware or software services that occur over the Internet through a third party. Qin, Hsu, and Stern (2016) refer to cloud computing as a form of computation "where the processing and applications mainly reside not on the user's computer or network, but rather on a remote server connected through an Internet connection" (p. 227). The business of cloud computing will continue to grow with an expected revenue of over $240 billion post-2020 (Dignan, 2011).

From a communication standpoint, cloud computing should have far reaching effects, especially on teams, teamwork, and group collaboration. As much of what we do – both as consumers and employees – moves into a cloud-based context, organizations should explore ways to improve collaboration services, specifically by focusing on team strategies surrounding communication, coordination, support, and effort

(Qin, Hsu, & Stern, 2016). As workplaces look to enhance the overall ethos of their organizations, as well as the efficiency of their operations, a thorough look at cloud computing will help increase productivity and overall group learning processes (Hadjileontiadou, Dias, Diniz, & Hadjileontiadis, 2015).

Learning and Development

Though learning and development are not new concepts, they will continue to be important focus areas for corporations and organizations in the future for several reasons. For one, learning and development initiatives continue to be linked to different business performance improvement (Chambel & Sobral, 2011). In addition, different knowledge areas, connected to learning and development, can serve as a key competitive advantage (Drucker, 2000). Generally, the area of learning and development has changed dramatically from one of individual, manual-driven instruction to high-tech interactive and experiential learning (Scurtu, 2015).

As a feature of future work, learning and development is expected to continue to evolve. It is expected that future learning and development will be even more mobile (Bonk, Kim, & Zong, 2005), will be focused not just on knowledge and behavior but also on affect (Barnett, 2012), and will be continually tech-driven and individualized to each learner (Davidson, 2009).

The post-pandemic world will continue to reveal new and necessary learning and development areas. However, it is expected that inclusivity and training in a potentially socially distanced or fully virtual workplace will continue to be a defining feature of learning and development moving forward (McBride, 2020).

Social, Emotional, and Physical Well-being

The other future work foci – automation, artificial intelligence, inclusive workplaces, and learning and development – are related, mostly, to job performance. However, as remote work continues to become more popular, work-life balance and other items related to general employee well-being have become a corporate focus for many organizations. Dodge, Daly, Huyton, and Sanders (2012) define well-being as "the balance point between an individual's resource pool and the challenges faced" (p. 222). Generally, though, well-being in the workplace is defined primarily positive. The fact that the employee is balanced, doing well, and is experiencing more positive emotions than negative emotions in the workplace (Aboobaker, Edward, & Zakkariya, 2019).

The work-from-home and work-from-anywhere culture has brought to the forefront a struggle for many employees, a healthy separation of work-life from home-life. Employees face a challenging context, a complex and blurred boundary between work and life (Hamilton Skurak, Malinen, Naswall, & Kuntz, 2018). In light of these challenges, organizations would do well to focus on programs and initiatives that emphasize employee well-being. As organizations focus on employee well-being, they may see increased productivity and performance (Peiró, Kozusznik, Rodríguez-Molina, & Tordera, 2019.

All told, these future work dynamics are exciting. While uncertainty definitely surrounds some of these developments, our workplaces will adapt. As the workplace of the future continues to evolve, relationships within organizations and initiatives that protect workplace culture should be a continued priority. Workplaces, then, need to appropriately respond not just to changing operations, like automation and artificial intelligence, but also to changing dynamics among employees. We explain how to seamlessly incorporate future generations into the workplace.

11.2 Integrating Future Generations and Creating a Culture of Generational Understanding

This book explores, primarily, relationships among Baby Boomers, Generation Xers, and Millennials. While some of our data and corresponding literature do address members of Generation Z, they are not a core focus of this text. Yet, members of Generation Z, and even those who come after members of Gen Z currently identified as Generation Alpha, should be considered when organizations develop a long-term culture of generational understanding.

Generation Z

Members of Generation Z, otherwise known as the iGeneration and post-Millennials, were born between 1997 and 2012 (Dimock, 2019). While not the only distinguishing factor of this generation, the iPhone launched in 2007, when the oldest Zers were 10. This generation's existence has been defined by technology, mobile devices, WIFI, and digital connectedness. This concept of digital native, truly connected from birth, continues to be a defining characteristic of Generation Z. In addition, diversity is an expectation, and Zers tend to be a little more pragmatic compared to their Millennial counterparts (Lanier, 2017). Finally, in the workplace, employers may have more success dealing with Generation Z through nondigital means because this generation tends to favor in-person communication with leaders (Schawbel, 2014). Generation Z will certainly continue to have a significant impact on 21st century workplaces, especially communication.

Generation Alpha

Generation Alpha currently are characterized by a birth year of 2013 or beyond. As you can imagine, little is known about this generation, and most conclusions are mere conjectures. However, we can project some distinguishing characteristics of this group. First, it is important to note that Alphas are, primarily, the children of Millennials. For those who routinely espouse Millennials as the source of all societal ills, this is bad news. We are already seeing projections of their sheer numbers (Alphas will account for roughly 11% of the global workforce by 2030), and their proclivities, as they are expected to delay life milestones, like marriage and child rearing, similar to previous generations. We may also be able to assume some other distinctions, specifically the influence of technology and constant connection and the formative experience of a global pandemic. While it would not be helpful at this time to determine how Generation Alpha will impact the workforce and workplace, it is worth noting that, like generations before them, they will bring changes, likely ones that are holistic and sweeping. This means workplaces should be continually adept at integrating new generations. The cycle will never end.

Integrating Generations

Pollak (2019) believes that a strong organizational culture is key to combating generational prejudice. She gives great advice when determining how to integrate generations in the workplace. Pollak (2019) says employers should

> Have as open conversation with people as possible with multigenerational employees at all levels about what flexibility means to them. Once you know what people want, you can prioritize what flexible options you can decide to offer. The data can show priority, then you can expand and regroup. (p. 230)

Pollak (2019) touches on a key concept of the future workplace: flexibility. While we can never truly combat or solve those challenges that stare down workplaces in a post-2020 world, we can solve some of those challenges by creating flexible work environments and by talking to our people about their needs and wants. In this vein, we can structure training to address automation and AI, create initiatives to encourage well-being, and deliver an environment where collaboration is not bound by time or geography. A workplace with different generations can be helpful as the organization navigates an ever-evolving set of challenges.

A Culture of Generational Understanding

To their credit, Lyons and LeBlanc (2019) view generational differences from the perspective of identity, not just demographic characteristics. This is helpful because, as they argue, it can reduce stereotypes and even intergenerational conflict (Lyons and LeBlanc, 2019). Ultimately, it is important to remember that every individual is not identified solely by their generational typology and, instead, individuals possess unique desires, especially regarding their careers. A study reiterates three distinct areas where generations differ: work ethic, managing change, and perception of organizational hierarchy (Glass, 2007). These realities may always exist; however, as younger generations become more ingrained in workplace dynamics, these differences may become less pronounced. Until then, organizations would do well to develop a corporate vision that enhances a welcoming understanding of all generations while placing people, no matter their age or generational affiliation, into an environment where they can succeed. Collaborative decision-making and training programs that focus on generational differences from a strengths perspective can also be helpful. Finally, organizations would do well to reconsider their communication efforts and ensure an environment of effective communication.

11.3 Communication "Next"

Effective communication must be a central component of the organization, now maybe more than ever. The influx of different generations into the workplace and the many demands of our communication time and energy necessitate a thorough yet flexible communication structure. Communication style differences exist between generations, and it is traditionally difficult to communicate across generational differences (Pollak, 2019).

The sheer avenues of communication have become multifaceted. Producing a strategy that effectively integrates all channels, both internal and external, should be a central feature of organizations as we continue to move forward toward the future of work. Leaders today should communicate early, communicate often, and communicate with transparency, when appropriate. Specifically, this means inclusive communication and engagement across levels and with employees of varying communication styles should be characteristic of leaders today. In addition, supervisors should actively look for ways to provide feedback, in person and virtually depending on the general tone and tenor of the organization.

The core communication trait needed today is adaptability. Leaders, managers, and employees should recognize that, like leadership, communication is not a one size fits all directive. Instead, people have individual communication preferences. Our current age of work should be approached with self-assessment and analysis. How do you prefer to have others communicate with you? Are you more formal or informal? Do you prefer email, phone, or face-to-face conversations? How do you prefer to receive (and offer) feedback? What level of detail do you desire when you receive messages? Asking yourself these questions and then asking these questions of your employees can help create a strategic communication climate.

As we navigate the future of work, we may also have an opportunity to return to the basics of communication. The channels will change, as they always have, but the core concepts remain the same. Effective communication continues to emphasize simplicity, clarity, and appropriate brevity. In addition, we will continue to see a shift in multimodal communication. Employee preferences for audio, video, and generally more visual communication may continue to increase because of the influx of younger generations but also because our habits and patterns societally are evolving. Organizations, then, would be wise to continue to communicate in ways that reflect broader cultural distinctives without forsaking core tenants.

11.4 Conclusion

There is much we do not know about the future of work, communication next steps, and how generational differences will continue to manifest and influence the organization. However, what we do know is that the workplace will continue to evolve. Organizations will change and adapt. Even in the midst of this evolution, though, employees will continue to search for jobs; companies will be defined by their organizational culture; organizational identification, mentorship, and supportive workplace communication will influence job satisfaction; and dissent and conflict will continue to be present in all organizations.

These organizational and communication variables are key features of work in the past, present, and future. As such, scholars would do well to continue to explore these variables as they relate to remote work, automation and artificial intelligence, well-being, training and development, and inclusivity. The findings presented in this volume provide a crucial starting point, a foundation for generational research at work, but there is more to be done. Our organizations are ripe for additional exploration.

References

Aboobaker, N., Edward, M., & Zakkariya, K. A. (2019). Workplace spirituality, employee wellbeing and intention to stay. *International Journal of Educational Management*, *33*, 28–42. doi:10.1108/IJEM-02-2018-004

Autor, D. H. (2015). Why are there still so many jobs? The history and future of workplace automation. *Journal of Economic Perspectives*, *29*, 3–30. doi:10.1257/jep.29.3.3

Barnett, R. (2012). Learning for an unknown future. *Higher Education Research and Development*, *31*, 65–77.

Bessen, J. (2019). Automation and jobs: When technology boosts employment. *Economic Policy*, *34*, 589–626.

Bonk, C., Kim, K., & Zong, T. (2005). Future directions of blended learning in higher education and workplace settings. In C. J. Bonk & C. R. Graham (Eds.), *Handbook of blended learning: Global perspectives, local designs* (pp. 22–25). Pfieffer Publishing.

Carmeli, A., Brueller, D., & Dutton, J. E. (2009). Learning behaviors in the workplace: The role of high-quality interpersonal relationships and psychological safety. *Systems Research and Behavioral Science: The Official Journal of the International Federation for Systems Research*, *26*, 81–98. https://doi.org/10.1002/sres.932

Chambel, M. J., & Sobral, F. (2011). Training is an investment with return in temporary workers: A social exchange perspective. *Career Development International*, *16*, 161–177.

Davidson, C. (2009). *Future of learning institutions in the digital age*. MacArthur.

Dignan, L. (2011). Cloud computing's real creative destruction may be the IT workforce. ZDNet. Retrieved from: https://www.zdnet.com/article/cloud-computings-real-creative-destruction-may-be-the-it-workforce/

Dimock, M. (2019). Defining generations: Where Millennials end and Generation Z begins. *Pew Research Center*, *17*, 1–7.

Dodge, R., Daly, A., Huyton, J., & Sanders, L. (2012). The challenge of defining wellbeing. *International Journal of Wellbeing*, *2*, 222–235.

Drucker, P. J. (2000). Knowledge work. *Executive Excellence, 17*, 11–12.

Gallaugher, J. (2014). *Information systems: A Manager's guide to harnessing technology*. Flat World Knowledge

Glass, A. (2007). Understanding generational differences for competitive success. *Industrial and Commercial Trainings*, *39*, 90–103.

Hadjileontiadou, S. J., Dias, S. B., Diniz, J. A., & Hadjileontiadis, L. J. (2015). Computer-supported collaborative learning: A holistic perspective. In L. A. Tomei (Ed.), *Fuzzy logic-based modeling in collaborative and blended learning, advances in educational technologies and instructional design* (pp. 51–88). IGI Global

Haenlein, M., & Kaplan, A. (2019). A brief history of artificial intelligence: On the past, present, and future of artificial intelligence. *California Management Review*, *6*, 5–14. https://doi.org/10.1177/0008125619864925

Hamilton Skurak, H., Malinen, S., Näswall, K., & Kuntz, J. C. (2018). Employee wellbeing: The role of psychological detachment on the relationship between engagement and work–life conflict. *Economic and Industrial Democracy*, 1–26. https://doi.org/10.1177%2F0143831X17750473

Howard, J. (2019). Artificial intelligence: Implications for the future of work. *American Journal of Industrial Medicine*, *62*, 917–926. https://doi.org/10.1002/ajim.23037

Kaplan, A., & Haenlein, M. (2019). Siri, Siri, in my hand: Who's the fairest in the land? On the interpretations, illustrations, and implications of artificial intelligence. *Business Horizons*, *62*, 15–25.

Lanier, K. (2017). 5 things HR professionals need to know about Generation Z: Thought leaders share their views on the HR profession and its direction for the future. *Strategic HR Review*, *16*, 288–290. https://doi.org/10.1108/SHR-08-2017-0051

Lyons, S. T., & LeBlanc, J. E. (2019). Generational identity in the workplace: Toward understanding and empathy. In R. J. Burke and A. M. Richardsen (Eds.), *Creating psychologically healthy workplaces* (pp. 270–291). Edward Elgar Publishing.

McBride, K. (2020, June 17). Training and development in a post-COVID-19 workplace. *Training Industry*. https://trainingindustry.com/blog/strategy-alignment-and-planning/training-and-development-in-a-post-covid-19-workplace/

Offerman, I. R., & Basford, T. E. (2014). Best practices and the changing role of human resources. In B. M. Ferdman & B. R. Deane (Eds.), *Diversity at work: The practice of inclusion* (pp. 580–592). Jossey-Bass.

Peiró, J. M., Kozusznik, M. W., Rodríguez-Molina, I., & Tordera, N. (2019). The happy-productive worker model and beyond: Patterns of wellbeing and performance at work. *International Journal of Environmental Research and Public Health*, *16*, 479–499. https://doi.org/10.3390/ijerph16030479

Peters, M. A. (2017). Technological unemployment: Educating for the fourth industrial revolution. *Educational Philosophy and Theory*, *49*. 1–6. doi:10.1080/00131857.2016.1177412

Pollak, L. (2019). *The remix: How to lead and succeed in the multigenerational workplace*. HarperCollins.

Qin, L., Hsu, J., & Stern, M. (2016). Evaluating the usage of cloud-based collaboration services through teamwork. *Journal of Education for Business, 91*, 227–235. https://doi.org/10.1080/08832323.2016.1170656

Rezai, M., Kolne, K., Bui, S., & Lindsay, S. (2020). Measures of workplace inclusion: A systematic review using the COSMIN methodology. *Journal of Occupational Rehabilitation, 30*, 1–35.

Sako, M. (2020). Artificial intelligence and the future of professional work: Considering the implications of the influence of artificial intelligence given previous industrial revolutions. *Communications of the ACM, 63*, 25–27. https://doi.org/10.1145/3382743

Sabharwal, M. (2014). Is diversity management sufficient? Organizational inclusion to further performance. *Public Personnel Management, 43*, 197–217. http://dx.doi.org/10.1177/0091026014522202

Schawbel, D. (2014, September 2). Gen Y and Gen Z global workplace expectations study. *Workplace Intelligence*. http://millennialbranding.com/2014/geny-genz-global-workplace-expectations-study

Scurtu, L. E. (2015). Knowledge and their shelf life in the business cycle. *Ecoforum 4*, 1–3.

Shore, L. M., Cleveland, J. N., & Sanchez, D. (2018). Inclusive workplaces: A review and model. *Human Resource Management Review, 28*(2), 176–189.

Wang, P. (2019). On Defining Artificial Intelligence. *Journal of Artificial General Intelligence, 10*, 1–37.

Winick, E. (2018, January 25). Every study we could find on what automation will do to jobs, in one chart. *MIT Technology Review*. https://www.technologyreview.com/2018/01/25/146020/every-study-we-could-find-on-what-automation-will-do-to-jobs-in-one-chart/

Index

Note: Page numbers in **bold** indicate tables in the text.

acculturation 70
administrative science 55
Afifi, W. 46
Amazon M-Turk 5
"American Dream" 17
articulated dissent 98
artificial intelligence (AI) 133
Asif, R. 95
assimilation 69–73; acculturation 70; dimensions 69; familiarity 69–70; generational difference 72–73; involvement 71; job competency 72; recognition 70–71; role negotiation 72
automation 132–133
Avery, G. C. 125

Baby Boomers 1, 6, 17, 19, 27, 43–44, 60, 67
Baker, E. 125
Bakker, A. B. 24
Bartik, A. W. 123
Basford, T. E. 134
Bessen, J. 133
best practices: ACORN your organization 34; audit, analyze, and adjust 34–35; be transparent 101; communicate clearly 75; communication 116; for conflict 100–102; create multimodal communication channels 34; for creating satisfying workplace 115–117; for crisis in modern multigenerational organization 100–102; culture 116; culture of advocacy 100–101; designate remote responsibilities 129; for dissent 100–102; enriching experiences 87–88; for internal communication 34–35; internal communication strategy 34; listening 88; mission-centric culture 75; platforms 129; purpose of remote work 128–129; start building relationships early and often 75; stress monitoring 88; technology 116–117; uncertainty management 88
bureaucratic organizations 54
Burgoon, J. 42
Burleson, B. 81
business communication 24
Bussin, M. H. R. 74

Campbell, S. 71
Career Builder 41
career development 42, 83–84
career mentoring 10
career-related issues 42
Cates, S. V. 17
certainty 92

challenges: organization 132–140; of remote and virtual work 125–126
Chandra, S. 92
change management 57
Chaudry, A. M. 95
Cheney, G. 67–68
Civil Rights Movement 19
Clampitt, P. G. 24
clarity 62
clarity of messaging 26
classical management approaches 54; *see also* administrative science
cloud-based collaboration 134–135
coaching 10
cohort 16
Cojanu, K. A. 17
Cold War 19
collegial–social 10
collegial–task 10
communicate clearly 75
communication: audit principles 25–26; business 24; clarity and 62; competency 27; corporate 24; dark side of 92–102; effective 138; effectiveness 24, 26–27; engagement and 25; exploring and evaluating in workplace 29–31; future nature of work 138–139; generational differences 86–87; hierarchical 28; infrastructure 16; management 24; in organization 3–4, 80–89; organizational 24; phenomenon 3; positive 80–89; preferences 26–27; style differences 138; supportive 80–82; systems 3; tools 25; workplaces 16; *see also* internal communication
Communication Satisfaction Scale **9**, 11, **12**, 111, **111**, 115
communicator reward value 46–47
competency 27
comprehensive multigenerational study 20
conflict 95–96; *see also* dissent
congruence 56
control strategies 11
corporate communication 24
COVID-19 121, 123, 133
Cravens, K. S. 100
Crawford, J. 125
Cullen, Z. B. 123
cultural identity 15
culture: assess 62–63; elements of 55; of generational understanding 138; leadership and 56; of organization 5, 53–63; *see also* organizational culture

Daly, A. 135
Davis, J. B. 67
Dell 85
Dencker, J. C. 95
development: career 42, 83–84; of communication systems 3; foundational 126; of generational differences 19; learning and 135; of multigenerational understanding 19–20; product 28; professional 84–85
disappointment 45–46
dissent 96–99; articulated 98; generational differences 99; latent 98
Dodge, R. 135
DOT Appropriations Act 126
Downs, C. W. 24, 106, 111
Drucker, P. 121

effective communication 138
Eisenberg, J. 125
emotional support 82
emotional well-being 135–136
employee: communication needs 26; engagement 25; preferences 26; recruitment 56
engagement 25
Engdahl, B. 86
Estee Lauder 60, 85
evolving organization 1–13; 21st century organization 3–5; demographics measures **7–8**; generational study 5–11; means 9; organizational change 2–3; overview 1–2; reliability coefficients **9**; scales and

subscales **9**, **12**; standard deviations **9**; survey and measures **7–8**
EVT *see* expectancy violations theory
excellence theory 28, 80
expectancy violations theory (EVT) 46–50
expectations, defined 42

Facebook 41
face-to-face interaction 122
face-to-face worker 122
familiarity 69–70
Fayol, H. 55
Flood, F. 123
formal means 39
foundational developments 126
future nature of work 132–140; artificial intelligence (AI) 133; automation 132–133; cloud-based collaboration 134–135; communication 138–139; future generations 136–138; inclusive work 134; learning and development 135; social, emotional, and physical well-being 135–136

Gallaugher, J. 134
General Electric 60, 85
Generation Alpha 137
generation, defined 15
Generation X. 1, 6, 17, 18–20, 44, 59–60, 67
Generation Y. 71
Generation Z/iGen 1, 5, 6, 17, 18–20, 27, 50, 72, 136–137
generational communication differences 24–35; communication effectiveness 26–27; effective internal communication, best practices 34–35; exploring and evaluating 29–31; internal communication 24–26; internal communication strategy 27–29; overview 24; *see also* internal communication
generational culture 17
generational data 43
generational differences 1, **32**, **33**, **43**; communication satisfaction and 111, **111**; conflict and 96, **97**; dissent and 99, **99**; efficiency 16; force managers 71; historical development of 19; indicator of 19; job satisfaction and 115; in job search strategies 40–41, **41**, **43**, **45**, **48**, **49**; mentoring and communication support and 86–87, **87**; in organizational assimilation index and subscales **73**; Organizational Communication Scale **30**; organizational culture and **61**, 61–62; remote work 127; in workplace 2, 16, 24–35
generational perspective, organization 15–21; generational differences 19; multigenerational understanding 19–20; multigenerational workplace 15–19; overview 15; *see also* multigenerational workplace
generational study 5–11; as detailed 5; goal of 5; as in-depth online survey 5; institutional research protocol approvals 5; open- and close-ended questions 5; student participants 5–6
Generational Theory 15
generational traits 16
Geyer, P. 115
gigification 122–123
Glaeser, E. L. 123
Golden, T. 114
Google Drive **127**
Google Hangouts 127
Gordon, J. 40
Granovetter, M. 39
Grant, C. A. 122

Hackman, J. 113
Hampden-Turner, C. 54
harassment 94
harmonious organizational culture 55

Harris, J. 86
Hazen, M. D. 106, 111
hierarchical communication 28
hierarchical sources 27
Higgins, C. A. 124
Hofstede, G. 54
Hollensbe, E. C. 95
home office working 122
Houston, A. 67
Howe, N. 15–16
Hsu, J. 134
Hulland, J. S. 124
human factor 2–3
Huyton, J. 135

identification 66–68; *see also* organizational identification
iGeneration 1, 5, 6, 17, 18–20, 27, 50, 72, 136–137
inclusive workplaces 134
Indeed.com 41
Industrial Revolution 132
ineffective communication in workplace 92–93; *see also* conflict; dissent; negative communication
informational support 82
internal communication: assessment of 25; best practices 34–35; domains 24; effective 24; hierarchical communication 28; history 24; measure of 26; overview of 24–26; platforms 25; promote and stimulate 24; strategies 25, 27–29; successful 29
international economic recession 4
International Women's Day 59
involvement 71

Jacobi, M. 84
job competency 72
job satisfaction 57–58, 112–115; *see also* workplaces
Job Satisfaction Scale **9**, 11, **12**
job search 4–5, 6; behavior 38–39; best practices 50; channels to 39; defined 38; direct application 40; disappointment 45–46; duration of **43**, 43–44; ease of 44–45; effort and ease of **45**; expectancy violations theory (EVT) 46–50, **48**; expectations for process 42–46; formal means 39; generational differences of 40–41, **41**; generational expectations of 38–50; individual differences 39; nonselection response **45**; overview 38; personal contacts 39; process 38–40; rejection 45–46; required effort 44–45; strategies 40–41, **41**; techniques 39; time commitment **43**, 43–44; websites 41
Jokisaari, M. 69
Joshi, A. 95
Jurkiewicz, C. L. 74

Kalla, H. K. 24
Kanungo, R. N. 67
Kassing, J. W. 96, 98–99
Kennelly, J. 128
Krishnan, A. 125
Kylili, A. 124

latent dissent 98
leadership 56
LeBlanc, J. E. 138
Lester, S. W. 16
Likert-style scale 10–11, 30; questions 6
LinkedIn 41
Luca, M. 123
Lwin, M. O. 92
Lyons, S. 42
Lyons, S. T. 138

management communication 24
Martin, C. A. 71
Martocchio, J. J. 95
mass media 27, 42
Masterson, S. S. 95
Mayer, Marissa 124
McDonald, P. 94
McGregor, D. 2–3
Mencl, J. 16
Mentoring and Communication Support Scale **9**, 10, **12**, 85, 86–**87**

mentorship 83–87
Metts, S. 46
Microsoft Teams 127
Mid-Atlantic United States 5
Mikkola, L. 82, 88
Millennial job seekers 42
Millennials 1, 5, 6, 16–17, 18, 20, 27, 58–59, 68, 71
mission-centric culture 75
Mitroff, I.I. 99
Mokhtarian, P. 121
Monster 41
motivating potential 113
multigenerational crisis response 99–100
multigenerational data 48, 58
multigenerational study 20
multigenerational understanding 19–20
multi-generational workforce 5, 29, 50
multigenerational workplace 17; communication effectiveness in 26–27; communication in 98; overview 15–17; supportive 87
Myers, K. K. 69, 70, 72

National Institute of Health (NIH) 121–122
nature of work 121–123
negative communication 93–94; *see also* conflict; dissent
Ng, E. 42
Nickson, D. 122
NIH *see* National Institute of Health
nonconfrontation strategies 11
Nurmi, J. E. 69

Obama, Barack 126
OCCI *see* Organizational Communication Conflict Instrument
OCS *see* Organizational Communication Scale
Odine, M 92
Oetzel, J. G. 69, 70, 72
Offerman, I. R. 134
Oishi, S. 100
Oldman, G. 113
Oliver, E. G. 100
O' Reilly, C. A. 29
organization: bureaucratic 54; challenges 132–140; communication differences in workplace 24–35; communication in 3–4, 80–89; culture of 5, 53–63; dark side of communication 92–102; dynamic organisms 1; evolutions 1–13; generational differences 1; generational perspective 15–21; human factor 2–3; identification 66–75; innovation 132–140; job search 38–50; mechanisms efficient machines 2; remote and virtual work 121–130; technology 2; workplace satisfaction 106–117
organizational assimilation *see* assimilation
Organizational Assimilation Index **9**, 10, **12**, 72, **73**
organizational change 1; global pandemic 4; international economic recession 4; investigation about 2–3; societal unrest in the United States 4
Organizational Communication Conflict Instrument (OCCI) **9**, 10–11, **12**, 96, **97**
Organizational Communication Scale (OCS) 6, 29, 30, **30**
organizational culture 16, 17, 53–63; age-inclusive 62–63; assess 63–64; best practices 62–63; defined 53–54; generational differences 61–62; generational perspectives on 58–60; history 54–55; overview 53; previous research and theory 55–58; *see also* harmonious organizational culture
Organizational Culture Survey 6–10, **9**, **12**, 61, **61**
organizational development 3
Organizational Dissent Scale **9**, 11, **12**, 98, 99, **99**
organizational identification 66–75; best practices 74–75;

generational difference 68; overview 66; workplace 66
Organizational Identification Scale **9**, 10, **12**, 68
organizational integration 112
organizational variables 4–5
Ozimek, A. 123

Patmos, A. K. 114
Pawlowski, S. D. 67
Peters, M. A. 132
Pettigrew, A. 57
Pettine, S. 17
Pew Research Center 40
physical well-being 135–136
Pitts, M. J. 114
Pollak, L. 26, 137
Pond, S. 115
positive communication 80–89
post-pandemic work 123
Pregnolato, M. 74
Procter & Gamble 85
product development 28
professional development 84–85
psychosocial support 84
public relations 80
Putnam, L. L. 96

Qin, L. 134

Rawlins, C. 58
realism 56
recognition 70–71
rejection 45–46
remote and virtual work 121–130; benefits of 124–125; best practices 128–129; challenges of 125–126; generational differences and **127**, 127–128; nature of work 121–123; platforms 129; purpose of 128–129; tools and platforms 126–127; types of 122; virtual effectiveness 124–126; work post-pandemic 123
retention 56, 74
Rieger, H. 57
Roberson, Q. M. 29
Roberts, K. H. 29
role negotiation 72
Roodt, G. 57

Sabharwal, M. 134
Saint-Gobain North America 60
Sanchez, P. 55
Sanders, L. 135
scales 9, **12**; Communication Satisfaction Scale **9**, 11, **12**, 111, **111**, 115; Job Satisfaction Scale **9**, 11, **12**; Mentoring and Communication Social Support Scale **9**, 86–87, **87**; Organizational Assimilation Index **9**, 10, **12**, 72, **73**; Organizational Communication Conflict Instrument **9**, 10–11, **12**, 96, **97**; Organizational Communication Scale 6, 29, 30, **30**; Organizational Culture Survey 6–10, **9**, **12**, 61, **61**; Organizational Dissent Scale **9**, 11, **12**, 98, 99, **99**; Organizational Identification Scale **9**, 10, **12**, 68
Schaufeli, W. B. 24
Schein, E. 56
Schlecter, A. F. 74
Schnackenberg, A. K. 100
Schullery, N. M. 15–16
Schweitzer, L. 42
scientific management cultures 54–55
securing employment 38
Sempane, M. 57
Serene 127
sexual revolution 19
Shakil, B. 97
Shou-Boon, S. 92
SHRM *see* Society for Human Resource Management
Siddiqui, D. A. 97
Siddons, S. 122
Silent Generation 6, 17
Simsek, Z. 114
Skype 127
Slack 3, 28, 126, **127**, 129
Smith, S. A. 114
social media/networks 28, 50
social norms 42
social support 83

social well-being 135–136
societal unrest in the United States 4
Society for Human Resource Management (SHRM) 113
solution-oriented strategies 11
Spark 127
Spurgeon, P. C. 122
Stanton, C. T. 123
Staples, D. S. 124
start building relationships early and often 75
Stern, M. 134
Stewart, J. S. 70, 100
strategies: control 11; internal communication 25, 27–29; job search 40–41, **41**; nonconfrontation 11; solution-oriented 11
Strauss, W. 15–16
supportive communication 80–82; defined 80–81; in workplace 82–83
supportive workplace 87–88
SWOT analysis 2

Telework Enhancement Act 126
teleworking 114
Theng, Y. 92
Theory X. 2–3
Theory Y. 2–3
time commitment 43–44
Time Warner 85
Toggl 127
Tolbize, A. 34
Tomlinson, E. C. 100
Traditionalists 17
Trello **127**
Trompenaars, F. 54
Tuikka, S. 93
Twenge, J. 71

uncivil relationships 94
United States 5; governmental actions in 126; Mid-Atlantic 5; societal unrest in the 4; unemployment in 121; working remotely in 126
Urick, M. J. 95

Veiga, J. 114

Wallace, L. M. 122
Wang, P. 133
well-being 135–136
Welsh, Jack 85
Whitworth, B. 29
WIFI 126
Wilson, C. E. 96
Winskowski, A. 86
work-from-home 124–126, 128, 132, 136; *see also* remote and virtual work
workplaces: communication 16; conflicts in 16; effectiveness 16; emotional support 82; generational differences in 16; generational similarities in 16; history in 19; inclusive 134; ineffective communication in 92–93; informational support 82; multigenerational 17; organizational identification 66; satisfaction 106–117; social support 83; supportive 87–88; supportive communication in 82–83; types of support in 82; uncertainty 82; *see also* multigenerational workplace
work post-pandemic 123

Yahoo 124

Zapier **127**
Zimmerman, K. 128
Zoom 127, **127**